THE
DANCING
UNIVERSE

UNDERSTANDING SCIENCE AND TECHNOLOGY SERIES

THE

FROM CREATION MYTHS

DANCING

TO THE BIG BANG

UNIVERSE

MARCELO GLEISER

DARTMOUTH COLLEGE PRESS

Hanover, New Hampshire

Published by University Press of New England / Hanover and London

To the memory of my parents

DARTMOUTH COLLEGE PRESS
Published by University Press of New England,
One Court Street, Lebanon, NH 03766
www.upne.com

Originally published in cloth in 1997 by Dutton, an imprint of
Dutton Signet, a member of Penguin Putnam, Inc.

First Dartmouth College Press/UPNE paperback edition 2005
ISBN 1–58465–466–X
Library of Congress Control Number 2004113702

Printed in the United States of America   5   4   3   2   1

The Library of Congress has cataloged the original edition as follows:
LIBRARY OF CONGRESS CATALOGING-IN-PUBLICATION DATA
Gleiser, Marcelo.
The dancing universe : from creation myths to the big bang /
Marcelo Gleiser.
p.   cm.
Includes bibliographical references and index.
ISBN 0–525–94112–6
1. Cosmology. I. Title.
QB981.G57   1997
523.1—dc21   97–17670

CONTENTS

ACKNOWLEDGMENTS

I would like to express my deep gratitude to the friends and colleagues who took their time to read the manuscript, helping me make it into a better text. Freeman Dyson's initial encouragement was really crucial to set me in the right direction. Rocky Kolb's influence as a friend, mentor (sorry Rocky, you are not that old, but . . .), and collaborator helped shape my career and this book. At Dartmouth, I learned a great deal from the comments and suggestions of Joseph Harris, Richard Kremer, and Kari McCadam. From physics to history to poetry, their contributions are reflected throughout this book. I would also like to thank my editor at Dutton, Deirdre Mullane, for her tireless efforts in making the original manuscript a more readable book; and my agents, Katinka Matson and John Brockman, for their advice and support.

To my children, Andrew, Eric, and Tali, for showing me how to always take a fresh look at the world and for injecting me with renewed energy every day. To Wendy, for her patience and understanding during the two long years it took me to make this book into a reality.

Finally, I would like to thank my father for teaching me to love the beauty around me and within other people.

Albert Einstein and Isidoro Kohn, his host during a visit to
Rio de Janeiro in 1925. This autographed photograph,
which has remained in my family since then, became almost
an icon as I devoted myself to the study of Einstein's contributions to our understanding of the Universe.

PREFACE TO THE 2005 EDITION

To write a book and see it printed is a privilege. To see it go through a new edition is an honor. I owe the final impetus to write *The Dancing Universe* to Freeman Dyson. When he was visiting Dartmouth in the fall of 1995, I asked if he thought a book dealing with creation myths and modern cosmology was something worth pursuing. At the time, I had written the first chapter about creation myths. After reading it he told me to go ahead but cautioned, "Don't add speculative modern ideas. It will shorten your book's usefulness, as most of them will probably be wrong." I did exactly that, resisting the temptation to write about inflationary cosmology, superstrings, extra dimensions, dark energy, and other juicy topics, even though they are an important part of my research (and some of them may be right after all). The book ends describing the first ideas concerning a quantum origin for the universe, a scientific version of the "creation out of nothing" mythic narrative from the seventies.

This new printing is faithful to the original version and to Dyson's advice. Covering over two thousand years of scientific ideas is enough of a challenge for a few hundred pages. But the effort is well worth it, as these ideas make for a fascinating story, the story of our quest for meaning in a universe that seems quite indifferent to our existence.

Since its publication in 1997, I have used *The Dancing Universe* as the main text in a large physics class for non-science majors. Perhaps the reason why the students have greatly enjoyed it is that it is *not* written as a textbook. I present science as a narrative, a human construction that attempts to reveal the intricacies of the world around us. As we change, so does the narrative. As the narrative changes, it transforms us. Science provides an ever-evolving view of Nature. The universe we live in today is very different from the one

Galileo or Plato lived in. It is also very different from the universe that people in the twenty-second century will inhabit. What is the same is our urge to understand, to make sense of our lives. As we inch forward, the universe keeps on dancing, carrying us in its arms. And with eyes wide open we watch in awe as we always have, and always will.

PREFACE

This book surveys the quest for understanding our origins and our place in this vast, mysterious Universe. For as long as history has been recorded humankind has asked the most fundamental question that can be asked, that of the origin of all things. As far as we know, every culture, past and present, has addressed this issue, arriving at various answers. From creation myths of prescientific cultures to modern cosmological theories, the question as to why there is something rather than nothing has inspired mythmakers and scientists, the religious and the atheist.

As we retrace the steps of this vast undertaking, we marvel at the manifold ways by which the human imagination has explained the mystery of creation. Beautiful metaphors and rich symbolism cross the boundaries between science and religion, expressing a true universality of human thought. This same universality, however, points to certain limitations of our imagination. The problem is that the sensorial perceptions and thought processes that we use to make sense of the world are bound by a polarized view of reality based on opposites such as day-night, cold-hot, female-male, matter-spirit, etc.; thus they can offer only a few logical ways to deal with what transcends this polarization, the Absolute from which all comes, be it conceived as God, a mythic "cosmic egg," or the laws of physics.

Although scientific and religious approaches to the question of the origin of the Universe have very little in common, certain ideas are bound to reappear, even if draped in completely different clothes. Thus, I begin with an analysis of creation myths of various cultures and close with a parallel discussion of modern scientific ideas concerning the origin of the Universe. By classifying creation myths and cosmological theories according to how they answer this question, I hope to make clear both their overlapping points and their marked differences.

In between, we will examine how our understanding of nature and the Universe as a whole evolved hand-in-hand with the development of physics, from its origins with the pre-Socratic philosophers of ancient Greece, all the way to the discovery of quantum mechanics and relativity during the first three decades of the twentieth century, the cornerstones of modern cosmological theories.

This book also looks at the people responsible for shaping our changing view of the Universe. Apart from explaining their science, I also explore the motivations, successes, and struggles of the many individuals that played a part in this long drama. As we will see, religion in many guises played (and plays!) a crucial role in the creative process of many scientists. Copernicus, the reclusive canon responsible for putting the Sun back at the center of the cosmos, was a reluctant revolutionary. Kepler, the man who told us that planets circle the Sun in elliptical orbits, was a rational mystic. Galileo, the man who first pointed the telescope at the stars, was a pious (but very ambitious) man, who believed he could single-handedly lead the Catholic Church into the new era of natural philosophy. Newton's universe was infinite, an embodiment of the infinite power of God. Einstein wrote that the devotion to science is the only truly religious activity in modern times.

By getting to know these scientists, we will not only better understand their science, but scientists in general. For this book is also about the way scientists think and feel about their work. There is no greater misconception of science than the widespread belief that scientists are cold and insensitive, a group of eccentrics pondering arcane questions no one else can understand. Science is often regarded as a purely rational activity in which objectivity reigns supreme as the sole avenue of knowledge. As a result, scientists are sometimes viewed as insensitive and narrow-minded, prone to strip away all the beauty of nature by approaching it mathematically. This labeling of scientists in general and physicists in particular has always bothered me. It misses the most important motivation for doing science, which is precisely our fascination with nature and its mysteries. Behind the scientist's complicated formulas, the tables of data obtained from experiments, and the technical jargon, you will find a person eagerly trying to transcend the immediate boundaries of life, driven by an unstoppable desire to reach some deeper truth. Seen this way, science is not so different from art, a process of self-discovery, as we try to capture the essence of ourselves and understand our place in the Universe. There is much more to physics than solving equations and interpreting data. I daresay that there is poetry in physics, that physics is a deeply human expression of our reverence for the beauty of nature.

I often think of a scientist writing about science for the general public as a translator, trying to convey the meaning of certain ideas in a language less precise than the original. There is no doubt that something will be lost in the translation; some meanings will be obscured by the simplification. In this book I will often appeal to your imagination, invoking images from everyday life that may help elucidate certain aspects of a theory. As with music, you don't have to know how to read a musical score to appreciate the beauty of a symphony. My hope is that the translation is still accurate enough that you can share in my delight at being a scientist, at being part of this wonderful quest for understanding.

PART I

O

BEGINNINGS

CREATION MYTHS

Every question possesses a power that does not lie in the answer.
— ELIE WIESEL

A t first sight, it may seem puzzling that a book written by a scientist about the evolution of cosmological thought should start with a chapter on the creation myths of ancient religions, but I have my reasons. First, in their variety these myths encompass all the logical answers we can give to the question of the origin of the Universe, including those found in modern theories of cosmology. This is not to say that modern science is simply rediscovering ancient wisdom, but that, when it comes to the question of the ultimate origin of all things, there is evidence for a true universality in human thought. The language is different, the symbols are different, but in their essence the ideas are the same.

There is, of course, a very big difference between a religious approach and a scientific approach to the origin of the Universe. Scientific theories are supposed to be testable and should be refuted if they don't correspond to reality. Even though we are at present far from being able to test any of the models describing the origin of the Universe, a given mathematical model will only be seriously considered by the scientific community if it can be tested against experiments. At present, all we can hope for is that these models will correspond to some important features of the observed Universe.

We will have much more to say about the promise and difficulties of these models later on. For now, what is important is to keep in mind that creation myths and cosmological models are both attempts to understand the existence of the Universe.

The second reason to start this discussion with creation myths is more subtle. These myths are essentially religious, an expression of awe as different cultures face the mystery of Creation. It is this very same awe that motivates much of the scientific creative process. My point is that *this awe itself* is more primitive than the particular way by which we choose to express it, be it in terms of organized religion or science. For most scientists, the drive comes from looking at nature as the ultimate challenge, and from a deep belief in the power of human reason to tackle it. Physics is a game played with nature, where we try to find out what's inside the box without being allowed to open it. In the words of Richard Feynman, expressed in his wonderful *Lectures on Physics*,

> Imagine that the world is something like a great chess game being played by the gods, and we are observers of the game. We do not know what the rules of the game are; all we are allowed to do is to watch the playing. Of course, if we watch long enough, we may eventually catch on to a few of the rules. The rules of the game is what we call fundamental physics.

You can interpret this text in two ways. One is to say that physics is plain fun; the other is to say that physics is more than fun, it is the language of the gods.

Mysticism, *if understood as the embodiment of our irresistible attraction to the unknown,* plays a fundamental role in the scientific creative process of many physicists, past and present. Neglecting this fact is closing our eyes to history and overlooking an essential aspect of science. In order to understand the roots of what might be called *rational mysticism,* we now turn our attention to the creation myths of prescientific civilizations.

The Nature of Creation Myths

Thousands of years ago, long before the body of knowledge we now call science existed, nature played a very different role in the fate of our species. Nature both gave food and warmth, and brought cold and destruction. It was benevolent and generous as well as ruthless and terrifying. Since humans could not control nature, they tried to appease it. Hoping that natural disasters such as volcanic eruptions, earthquakes, and hurricanes would not destroy their homes or kill their animals, and that their harvest would be plentiful, people worshiped and idolized nature. The ways in which this occurred varied from culture to culture, conditioned by such factors as the degree of isolation of the group, its geographical location, and its technological sophistication.

In certain cultures a host of gods controlled (or even personified) several natural phenomena, while in others nature itself was a god, such as the Mother Goddess. Rituals and offerings were meant to appease the gods (or God or Goddess) in an effort to insure the survival of the group. Through this relationship with the gods, humans were trying to order their existence, giving meaning to phenomena that were unpredictable and mysterious. On the other hand, the relationship with the gods also had a more immediate social function, serving to impose moral values that were crucial to the cohesion of the group.

This religious relationship with nature went beyond the concrete needs of the group's well-being, also encompassing more metaphysical needs. A typical example is the role death plays in different religions. In some, death is only a gateway into another life, a passage from one existence into another in a cycle that repeats itself eternally. In others, death represents an ascension into an Absolute Reality, the promise of a deserved existence in Paradise after the many tribulations of life. In all cultures the search for a religious understanding of death fulfills the need to cope with that which is unpredictable and unexplainable. To the believer, this offers comfort when facing the loss of a loved one, and the certainty that his or her own death is not the end of existence. To the unbeliever, science may provide some comfort. In the words of Harvard physicist Sheldon Glashow: "Perhaps, by understanding science, we can more easily face our own mortality and that of our species and our planet." Science as an antidote for fear.

Another subject in which religion has traditionally played an important role is the origin of the Universe, or the question of Creation. This is perhaps the most fundamental question we can ask about our existence, so much so that I have come to think of it as "the Question." After all, you are reading this book because somehow the Universe allows for intelligent life to develop, so that (at least) one species, inhabiting a small planet orbiting a small star in one of the hundreds of billions of galaxies in the Universe, can ask questions about its origin. As we ask about our origin, or the origin of life, we are implicitly asking about the origin of the Universe, "the origin of origins." It is then no surprise that modern cosmology exerts such fascination these days. Due to its very nature, science should offer universal answers, independent of a particular religious or moral point of view. As scientists address the origin of the Universe, they play, at least to the society at large, the role of universal mythmakers, transcending boundaries of race and creed.

As soon as we start reflecting about the origin of the Universe, we realize that we must confront some very fundamental problems. How can we comprehend the origin of "everything"? If we assume that "something" created "everything," we fall into an infinite regression; who, then, created the "something" that created the "everything"? If we say that "nothing" existed before anything, we are implicitly assuming the existence of the "everything" that is its opposite. The absurd limit of this is seen in the dialogue between Alice and the Red King, in Lewis Carroll's *Through the Looking Glass*, when the Red King asks Alice: "What do you see?," to which Alice answers "Nothing." The King, impressed, comments, "What good eyes you have!"

When trying to understand the Universe as a whole, we are limited by our "insider's" perspective, somewhat like a clever fish trying to describe the ocean as a whole. This is true both in religion and in science. It is particularly acute in the realm of quantum cosmology, where quantum mechanics is applied to describe the origin of the Universe.*

---

*The word *quantum* refers to the physics of the very small, the world of atoms and subatomic particles.

In traditional quantum mechanics, the observer has a privileged role. The observer's presence, his or her perspective, is somehow responsible for the outcome of a given experiment. In order to apply quantum mechanics to the Universe as a whole, the role of the observer has to be modified, roughly speaking, because no one is there to measure anything. After all, measurement in principle assumes a separation between the observer and the observed. But we can't step out of the Universe to measure its properties. Our reasoning stumbles against an apparently insurmountable barrier, which has its origins in the way we think and behave socially, the polarization inherent in our perception of reality. Being able to distinguish between opposites is a basic requirement for ordering the world around us. Our lives and choices are routinely based in opposites such as night or day, cold or hot, guilty or innocent, dead or alive, rich or poor. Without such distinctions our values would not make sense, our agriculture would not be productive, and our species would probably not survive.

The problem is that we pay a price for our survival. Questions that transcend the distinction between opposites remain unanswered. At least they remain without an answer that we can consider logical. But this does not detain our curiosity. On the contrary, what is fascinating is that in *all* cultures "the Question" was asked. The need to understand our origin and that of the entire Universe, that is, the problem of Creation, is inherently human, traversing barriers in time and space. It was with us then, as we gathered inside the cave during a storm, and it is with us now, if we only find the time to reflect about who we are.

Once we ask about the origin of the Universe, finding an answer becomes very tempting. The road one chooses to find a solution depends, of course, on who is asking the question. A religious person will look for answers within the framework of a religion, which may be an established religion or a collection of personal beliefs. An atheist may look for an answer suggested by scientific explorations. Religious or not, I am sure that most people settle upon some explanation. The common vehicle found by prescientific cultures to address this problem was the myth. Myths are stories that bring meaning and order to our existence, serving as a valuable probe into the worldview and values of a particular culture.

The strength of a myth lies not in its being right or wrong, but in its being satisfying. In fact, being right or wrong is quite irrelevant. This is never more true than with *creation, or cosmogonical—* from the Greek *kosmos + gonos (origin)—myths.* Clearly, when a given culture tries to find an explanation for the origin of "everything," it must use language that is essentially metaphorical, based on symbols relevant to that particular culture. Metaphors are also common in science, especially science that explores phenomena beyond our sensorial experience, such as the world of the very small and the very fast, the atomic and subatomic realm.

This explains why myths of a certain culture may appear totally meaningless to another. In fact, a common mistake when examining myths of other cultures is to interpret them with symbols and values of our own culture. An even worse mistake is trying to interpret a myth scientifically, or giving a myth scientific value. A given myth is properly understood only within the context of the culture that created it. For example, the Assyrian myth "Another Version of the Creation of Man" (*c.* 800 B.C.) starts with five great gods, Anu, Enlil, Shamash, Ea, and Anunnak, discussing the progress of creation. Without knowing the importance of these deities to the Assyrian people, the image of five gods talking up in their heaven seems simplistic. However, once we realize what each god represents, the myth makes much more sense. Anu symbolizes the sky or air, Enlil the Earth, Shamash the Sun or fire, Ea water, and Anunnak destiny. For the Assyrians, creation was a process in which the four elements and time unfold together. Their religion was firmly rooted in rituals that celebrate the power of nature; the people's mission was to maintain and increase the power and fertility of the Earth.

Due to their profound nature, creation myths portray more effectively than any other kind of myth how a given culture perceives and organizes the world around it. In the myths selected in this book, the goal is not to offer a detailed analysis of creation myths using tools from cultural anthropology, but to present the answers found by different cultures to "the Question." Within this narrower focus, we will see that the creation myths chosen exhaust all the possible answers to the question of the origin of the Universe. That is, after they are stripped of their often colorful (and beautiful) symbolism the myths can be grouped in accordance to the way in which they explain the Creation (or its absence!). Later on, when we discuss twentieth-century cosmology and its attempts to describe the origin and evolution of the Universe, we will find some traces of these ancient ideas, distant memories perhaps, lingering evidence of the universality of human thought.

A CLASSIFICATION OF CREATION MYTHS

As we have seen, the fundamental problem in searching for the origin of "everything" is the limitation imposed by our bipolar perception of reality. The process or entity (or entities) responsible for the Creation must create the entire panoply of opposites, being thus large enough to encompass all of them. The solution to this problem is, in most cultures, a religious one. In general, the existence of an absolute reality, or simply Absolute, is assumed, which not only includes but transcends all opposites. The Absolute is the central element in all credos, giving creation myths a deeply religious character. In most cases the Absolute, which embodies the synthesis of all opposites, exists by itself, independently of the existence of a Universe. It has no origin, being beyond any cause-effect relationship. This Absolute could be God, or the realm of many gods, or the primordial Chaos, or even the Void, the non-being. On the other hand, we live in our polarized reality from where we try to comprehend the essence of this Absolute. The bridge connecting the all-encompassing Absolute to our structured reality is the creation myth. Through their creation myths religions proclaim their validity, relating the comprehensible to the incomprehensible. They portray the origin of the Universe as a consequence of the disintegration of the unity inherent in the Absolute into the polarized reality of the world.

What are, then, the various answers offered by different cultures to "the Question"? One example of the strong symbolism used in creation myths is found in the following myth of the Hopi Indians of the United States:

First, they say, there was only the Creator, Taiowa. All else was endless space. There was no beginning and no end, no time, no shape, no life. Just an immeasurable void that had its beginning and end, time, shape, and life in the mind of Taiowa the Creator.

Then he, the infinite, conceived the finite. First he created So-tuknang to make it manifest, saying to him, "I have created you, the first power and instrument as a person, to carry out my plan for life in endless space. I am your Uncle. You are my Nephew. Go now and lay out these universes in proper order so they may work harmoniously with one another according to my plan."

Sotuknang did as he was commanded. From endless space he gathered that which was to be manifest as solid substance, molded it into forms.

Here, as in many myths, the Infinite creates the finite, giving concrete shape to matter, molding it into forms. Taiowa represents the Absolute, which is omnipresent (present simultaneously in all places), omniscient (has absolute knowledge), and omnipotent (has infinite power). The Universe is created through the action of a "Positive Being," at a particular moment. That Creation occurs at a specific moment in time implies that the Universe has a finite age.

In other myths, a very different viewpoint of the role of time is assumed. The Universe was not created at an exact moment, but has existed and will exist forever. A well-known example comes from Hindu mythology. Here time has a cyclic nature: Creation repeats itself eternally in a cycle of creation and destruction symbolized by the rhythmic dance of the god Shiva.

In the night of Brahman (the ultimate reality, the inner essence of all things, infinite and incomprehensible), nature is inert, and cannot dance till Shiva wills it. He rises from His rapture and, dancing, sends through inert matter pulsing waves of awakening sound, and lo! matter also dances, appearing as a glory round about Him. Dancing, He sustains its manifold phenomena. In the fullness of time, still dancing, He destroys all forms and names by fire and gives new rest.

The dance of Shiva symbolizes all that is cyclic in the Universe, including its own evolution. Through His dance the God creates the Universe and its material contents, maintaining it during its existence until the time comes for its destruction. This cycle repeats itself throughout eternity, without beginning or end. To the Hindus, existence manifests itself through the dynamic interplay of opposites, life and death, creation and destruction, mirroring the many cyclic natural phenomena that rule our lives. The dance of Shiva symbolizes not only the rhythmic nature of time, but also the ephemeral nature of life, helping believers face their own mortality.

As we examine the evolution of cosmological thought and the role rational mysticism played and plays in the creative process of some of the greatest physicists of all time, we will seek in these creation myths the basic ideas about Creation lying behind their rich symbolism, the *answers* provided by the various creation myths to the problem of Creation, leaving aside details of the cultures that generated them.

The various creation myths can be separated into two main groups, according to how they address the question of "the Beginning." Either the myths assume that there was a specific beginning, a moment of Creation (though not necessarily an *act* of Creation), as in the example from the Hopi Indians, or they assume that the Universe existed forever, as in the case of the dancing Shiva. In the first case the Universe has a finite age, while in the second case its age is infinite. The reader may argue that in the case of a cyclic Universe each cycle begins with a Creation, yet the effect of having an infinite number of cycles is that one cannot speak of a "Beginning," but must instead speak of infinite beginnings, none more relevant than the other.

Myths that assume a single moment of Creation can be called simply "Creation Myths." Myths in which the Universe is eternal, or created and destroyed an infinite number of times, can be called "No-Creation Myths." Within each group there are subgroups, defined according to the way they portray the process responsible for the existence of the Universe.

"Creation Myths" can be subdivided into three groups, depending on the agent of Creation. The Universe may be the creation of a God, or Goddess, or gods, which can be called "Positive Being." The Universe may appear from Nothing, an absolute emptiness, which can be called "Negative Being" or "Non-Being." Or the Universe may appear from a primordial Chaos, where being and non-being

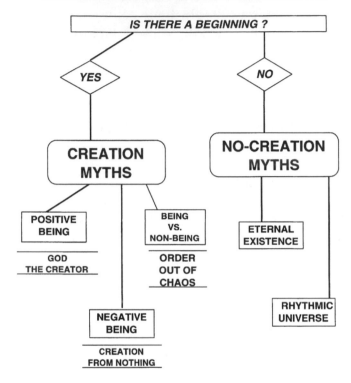

A CLASSIFICATION
OF
COSMOGONICAL MYTHS

IS THERE A BEGINNING ?

YES

NO

**CREATION
MYTHS**

**NO-CREATION
MYTHS**

POSITIVE
BEING

BEING
VS.
NON-BEING

ETERNAL
EXISTENCE

GOD
THE CREATOR

ORDER
OUT OF
CHAOS

NEGATIVE
BEING

RHYTHMIC
UNIVERSE

CREATION
FROM NOTHING

Different mythological responses to "the Question," the origin of the Universe

coexist. In this case, Creation occurs when matter coalesces from the tension between being and non-being and proceeds to differentiate itself into the various shapes of nature and life. All three possibilities share the same idea of a moment of creation, an instant when time itself comes to be. We can think of time as a straight line starting at t=0.

"No-Creation Myths" can be subdivided into two groups. Since there is no unique moment of creation, the only possibilities are for either a Universe that existed and will exist for all eternity, never created and never destroyed, or for a Universe that is continually being created and destroyed, in a cycle that repeats itself forever. In the first case, we can visualize time as a straight line that originates infinitely far away from where we are now. Thus, all points on the line are equivalent, and the beginning of time becomes subjective, an arbitrary starting point. In the second case, we can visualize time as a circle, so that it always comes back to its starting point. Again, there is no special point that we can identify with the beginning of time itself.

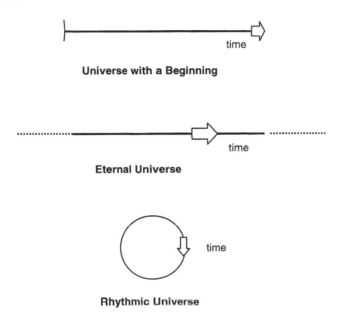

A pictorial representation of time in several myths

Types of Creation Myths

Myths that assume a beginning are by far the most common type of creation myths, particularly those that invoke a "Positive Being" in the role of the Creator. For Western cultures, the best-known creation myth is found in Genesis (*c.* 400 B.C.):

> In the Beginning God created the heavens and the earth.
> The earth was without form and void, and darkness was upon

the face of the deep; and the Spirit of God was moving over the face of the waters.

And God said, "Let there be light"; and there was light.

And God saw that light was good; and God separated the light from the darkness.

God called the light Day, and the darkness He called Night. And there was evening and there was morning, one day.

Here God, the Absolute, creates the world through His word. He orders light into existence, and separates it from darkness, its opposite. Creation occurs in two steps, the coming into being of something (in this case light), which is followed by its being separated from its opposite. This is an example of what some authors call "God the Thinker," as Creation is in a sense a rational act, expressed by words. The same idea appears in many other myths, as in the Assyrian myth with five gods discussed above and the Mayan Popol Vuh, where the gods Tepev and Guevmatz, symbolizing sun-fire in the middle of the dark waters, say together "Let the emptiness be filled!"

Another example of a Positive Being is that of "God the Great Organizer," who creates the Universe by bringing order into the primordial Chaos. A beautiful example appears in the *Metamorphoses*, written in A.D. 8 by the Roman poet Ovid, a rare expression of Roman concern with such questions.

> Before land was and sea—before air and sky
> Arched over all, all Nature was all Chaos,
> The rounded body of all things in one,
> The living elements at war with lifelessness;
> No God, no Titan shone from sky or sea,
> No Moon, no Phoebe outgrew slanted horns
> And walked the night, nor was Earth poised in air.
> No wife of Ocean reached her glittering arms
> Into the farthest shores of reef and sand.
> Earth, Air, Water heaved and turned in darkness,
> No living creatures knew that land, that sea
> Where heat fell against cold, cold against heat—
> Roughness at war with smooth and wet with drought.

Things that gave way entered unyielding masses,
Heaviness fell into things that had no weight.

Then God or Nature calmed the elements:
Land fell away from sky and sea from land,
And aether drew away from cloud and rain.
As God unlocked all elemental things,
Fire climbed celestial vaults, air followed it
To float in heavens below; and earth which carried
All heavier things with it dropped under air;
Water fell farthest, embracing shores and islands.

When God, whichever God he was, created
The universe we know, he made of earth
A turning sphere so delicately poised
That water flowed in waves beneath the wind . . .

Chaos here does not represent destruction or disorder, but the potential for the existence of all opposites, though their existence is not yet manifest: "The living elements at war with lifelessness . . . Where heat fell against cold, cold against heat—Roughness at war with smooth and wet with drought. . . ." And then God, of unexplainable origin, appears and organizes Chaos, separating the opposites and arranging the basic elements (fire, air, earth, and water) in their places, in accordance with Aristotle's dictums.

Still within the Positive Being subgroup, some myths depict God as a craftsman, as in the Hopi myth seen above, or in the second myth of Genesis, wherein God forms Adam from dust and the breath of life. Others use a procreation metaphor that appears in many different guises: a Mother Goddess who literally gives birth to the Earth or to lower gods who fashion the Earth; male and female gods embrace, giving birth to the Universe as a result; or a single God who either creates a mate or has a feminine dimension, with which he creates the world. Finally, a Positive Being may invoke a divine sacrifice to give rise to Creation. The Absolute God dies, or mutilates him- or herself to give rise to Creation. An example can be found in one of the several versions of the Chinese myth of P'an Ku (*c.* third century A.D.):

> The world was never finished until P'an Ku died. Only his death could perfect the Universe: From his skull was shaped the dome of the sky, and from his flesh was formed the soil of the fields; from his bones came the rocks, from blood the rivers and seas; from his hair came all vegetation. His breath was the wind, his voice made thunder; his right eye became the Moon, his left eye the Sun. From his saliva or sweat came rain. And from the vermin that covered his body came forth mankind.

A second kind of Creation Myth assumes that nothing existed prior to Creation. There is no God or gods, just emptiness, or Negative Being. Creation appears from this nothingness, without much reason as to why it happens. An example comes from the Hindu Upanishads:

> In the beginning this [world] was nonexistent. It became existent. It turned into an egg. The egg lay for the period of a year. Then it broke open. Of the two halves of the eggshell, one half was of silver, the other of gold. That which was of silver became the Earth; that which was of gold, Heaven. What was the thick membrane [of the white] became the mountains; the thin membrane [of the yolk], the mist and the clouds. The veins became the rivers; the fluid in the bladder, the ocean. And what was born of it was yonder Aditya, the Sun. When it was born shouts of "Hurrah" arose, together with all beings and all objects of desire. Therefore, at its rise and its every return, shouts of "Hurrah!" together with all beings and all objects of desire arise.

The theme of the cosmic egg is very common in creation myths. In one version of the P'an Ku myth, he appears from inside a cosmic egg. The interesting aspect of this particular myth is that the egg appears from nothing, and Creation seems to happen spontaneously, without the intervention of a divine being, by the dissociation of the cosmic egg. The egg in this myth plays the same role that P'an Ku plays in the myth related above, that of the source of all things. But here there is no sense of divine sacrifice for the Creation, but simply the familiar model of an egg hatching. There is no attempt to explain where the egg comes from; it simply "became" out of a Universe that also came to be.

Another example of creation out of nothing comes from the Maori natives of New Zealand; in the myth, "The Creation":

> From nothing the begetting,
> From nothing the increase,
> From nothing the abundance,
> The power of increasing the living breath;
> It dwelt with the empty space,
> And produced the atmosphere which is above us,
> The atmosphere which floats above the Earth;
> The great firmament above us dwelt with the early dawn,
> And the Moon sprung forth;
> The atmosphere above us dwelt with the heat,
> And then proceeded the Sun;
> They were thrown up above,
> As the chief eyes of Heaven:
> Then the heavens became light,
> The early dawn, the early day, the midday.
> The blaze of day from the sky.

There is no Being responsible for this Creation, which appears from nothing, simply due to an inexorable urge to be.

A final kind of Creation Myth states that Creation is the result of the tension between Being and Non-Being, both coexisting in the primordial Chaos. Unlike Ovid's cosmogony, though, there is no God taking charge of Creation; it just happens as order coalesces out of Chaos, through the dynamic interplay of opposite tensions. In modern scientific jargon, we could say that complexity emerges from disorder as a spontaneous manifestation of self-organization. This idea is clearly expressed in a Taoist myth from before 200 B.C.:

> In the beginning there was chaos. Out of it came pure light and built the sky. The heavy dimness, however, moved and formed the

earth from itself. Sky and earth brought forth the ten thousand creations, the beginning, having growth and increase, and all of them take the sky and earth as their mode. The roots of Yang and Yin—the male and female principle—also began in sky and earth. Yang and Yin became mixed, the five elements separated themselves from it and a man was formed.

The opposites are represented by Yin and Yang, with Yin representing darkness, passivity, and weakness, while Yang represents brightness, activity, and strength. Creation results from the interplay of opposites.

TYPES OF "NO-CREATION" MYTHS

Let's now turn briefly to the "No-Creation" myths, such as the rhythmic creation of the Universe through the dance of Shiva that we've already seen. Another example of a myth depicting an eternal Universe comes from Jinasena (c. A.D. 900), a teacher of Jainism, a religious faith of India said to have been started by Mahavira, a contemporary of Buddha in the sixth century B.C. Jinasena completely rejects the idea of Creation in a chain of arguments that I find quite lucid, despite its ill-tempered tone:

Some foolish men declare that Creator made the world. The doctrine that the world was created is ill-advised, and should be rejected.

If God created the world, where was he before creation? If you say he was transcendent then, and needed no support, where is he now?

No single being had the skill to make this world—For how can an immaterial god create that which is material?

How could God have made the world without any raw material?
If you say he made this first, and then the world, you are faced with an endless regression.
If you declare that this raw material rose naturally, you fall into another fallacy,
For the whole Universe might thus have been its own creator, and have arisen equally naturally.

If God created the world by an act of his own will, without any raw
material,
Then it is just his will and nothing else—and who will believe this
silly stuff?
If he is ever-perfect and complete, how could the will to create have
arisen in him?
If, on the other hand, he is not perfect, he could no more have created the Universe than a potter could. . . .

If he is perfect, what advantage would he gain by creating the
Universe?
If you say that he created to no purpose, because it was his nature
to do so, then God is pointless.
If he created in some kind of sport, it was the sport of a foolish
child, leading to trouble. . . .

Thus the doctrine that the world was created by God makes no
sense at all.
Good men should combat the believer in divine creation, mad-
dened by an evil doctrine.

Know that the world is uncreated, as time itself is, without begin-
ning and end,
And is based on the principles, life, and the rest.
Uncreated and indestructible, it endures under the compulsion of
its own nature,
Divided into three sections—hell, earth, and heaven.

In this text the Universe is uncreated and indestructible, maintained and changing according to natural principles. For the Jains,
the point was to release the soul from the Hindu eternal cycle of
transmigration into an eternal state of omniscient inactivity.

Thus, we have seen examples of all five types of cosmogonical
myths that I have identified in the diagram on page 12, which I propose exhaust the ways by which creation myths make sense of the
problem of the origin of all things. However, there is one more alternative, which is to admit that the problem of the origin of all
things is not accessible to human understanding and will necessarily

remain a mystery. Thought is derivative of being, and is thus insufficient to answer questions about the origin of being. This view is clearly stated in a Hindu Veda, *c.* 1200 B.C.:

> When neither Being nor Non-Being was
> Nor atmosphere, nor firmament, nor what is beyond.
> What did it encompass? Where? In whose protection?
> What was water, the deep, the unfathomable?
> Neither death nor immortality was there then,
> No sign of night or day.
> That One breathed, windless, by its own energy:
> Nought else existed then.
> In the beginning was darkness swathed in darkness;
> All this was but unmanifested water.
> Whatever was, the One, coming into being,
> Hidden by the Void,
> Was generated by the power of heat.
> In the beginning this One evolved,
> Became desire, the first seed of mind.
> Wise seers, searching within their hearts
> Found the bond of Being in Non-Being.
> Was there a below? Was there an above? . . .
>
> Who knows truly? Who can declare it?
> Whence it was born, whence is this emanation.
> By the emanation of this the gods only later came to be.
> Who knows whence it has arisen? . . .
>
> Only He who is its overseer in highest heaven knows.
> He only knows, or perhaps He does not know!

In this account, there is a being responsible for Creation, but the speaker is skeptical of its nature and essence. The lesser gods are ignorant of the true purpose of Creation, and even the One may not know. There is no clear answer, for the true nature of Creation is incomprehensible.

So, we have seen how some prescientific cultures have explained the problem of Creation. Next I will trace the evolution of Western science, from the pre-Socratic philosophers all the way to twentieth-century physics. As we go along, I will emphasize how our evolving understanding of natural phenomena changed the way we picture the Universe and our place in it. The gradual development of a rational approach to nature created an alternative way to address some of the mysteries that previously were the exclusive province of religion.

As more and more natural phenomena were understood by science, religion slowly gave up its absolute control over the ways of the world, becoming mostly concerned with the ways of the spirit. This "parting of the ways" between science and religion did not come without drama. In fact, this drama is still pretty much alive today, due to unfortunate misuses of both science and religion. As I retell some chapters of this long story, we will encounter such figures as Galileo, Newton, and Einstein, exploring not only their science but also their motives and beliefs, which are so often forgotten. If I am successful, by the end of this book you will regard the popular image of the creative scientist as a cold rationalist as complete nonsense (if you don't already!), like most generalizations. If I am very successful, by the end of this book the practice of science will mean something quite different from the mere exploration of natural phenomena. Science will come to represent the focus of very human aspirations, of the need to understand our origin and our destiny, as we face the vastness of this mysterious Universe.

The science-religion debate is usually restricted to how compatible the two are: Can a person approach the world scientifically and still be religious? I think the answer is an obvious yes, as long as the inquiries don't interfere with each other the wrong way. Scientists should not apply science abusively to situations where it is still clearly speculative, and claim they understand questions of theological nature. Religious people should not try to interpret religious texts scientifically, as they were not written for that purpose. To me, what is truly fascinating is that both science and religion express our reverence for nature. Their complementarity is manifest in the essentially religious motivation of many of the scientific heroes of every era. The awe that moved them, and that moves me into being a scientist today, is in essence the same awe that moved the mythmakers of times past. As we, in the silent confines of our offices, address the most fundamental questions about the Universe scientifically, we can hear, under the monotonous humming of our computers, the chants of our ancestors echoing through time, inviting us to sing along.

2

THE GREEKS

The real constitution of things is accustomed to hide itself.
— HERACLITUS

In his dedication of *Advancement of Learning* (1605) to James I, Sir Francis Bacon declared that "of all the persons living that I have known, your Majesty were the best instance to make a man of Plato's opinion, that all knowledge is but remembrance." Although Plato may have meant this to express his belief in the immortality of the soul, and Bacon used it as a ploy to obtain the King's favors (which worked quite well), we can see in it the tremendous importance that Greek thinking has had in the development of Western culture. From architecture to sculpture to theater to philosophy, the Greeks provided lasting foundations.

After defeating the Persians in a series of conflicts during the first decades of the fifth century B.C., Greece entered a century and a half of great splendor, inspired by the leadership of Pericles, who ruled Athens for over thirty years (461–429 B.C.). Not even the bitter disputes for power among Athens, Sparta, and other states that triggered the Peloponnesian War from 431 to 404 B.C. detracted from the remarkable level of sophistication achieved in this era. In the words of H. G. Wells, "During this period the thought and the creative and artistic impulse of the Greeks rose to levels that made their achievement a lamp to mankind for all the rest of history."

That this lamp kept burning through later centuries of religious intolerance and endemic warfare is our debt to those who believed knowledge to be above the blindness caused by greed and fear.

The first sparks to light the lamp came from the epic poems of the legendary "blind bard" Homer, the *Iliad* and the *Odyssey*, probably written during the eighth century B.C. By that time there were Greek settlements spread throughout the Mediterranean, from southern Italy and Sicily to the Black Sea and Asia Minor, now Turkey. These epics, together with the Olympic Games, offered a common cultural reference to the many small villages separated by ocean, mountains, and even race. Based on stories about the barbarian conquests of the Greeks around the time of the Trojan War (twelfth century B.C.), the poems served to link the many tribes by language, ancestry, and values. The Universe was described as an oyster-shaped Earth (like the shield of the great hero Achilles), surrounded by an ocean-river, as in earlier Babylonian cosmologies. In the *Odyssey*, Homer describes the starry heaven as made of bronze or iron, supported by pillars. There are references to several constellations, like Orion and the Pleiades, and to the phases of the Moon.

These concepts, however, certainly do not compare with the level of sophistication achieved by Babylonian astronomers, who a thousand years earlier had compiled detailed tables of planetary motion extending over decades. The Venus tablets of Ammizaduga (*c.* 1580 B.C.) detailed the risings and settings of Venus over a period of twenty-one years. These tablets were calendars used for organizing group activities, such as farming and religious ceremonies, and for astrological forecasting. But, notwithstanding the remarkable achievements of the Babylonians in astronomy, their Universe did not differ from Homer's in that it was also populated and ruled by gods. The Babylonian creation myth, the *Enuma elish*, or "When above," describes the origin of the Universe and the subsequent appearance of order as the work of several gods. There was no attempt to understand the cause of the heavenly motions, as mythic explanations were perfectly satisfactory. This situation would change, at least temporarily, two centuries after Homer, during the so-called pre-Socratic period of Greek philosophy, spanning roughly one century of Greek thought from the early sixth century to the birth of Socrates *c.* 470 B.C. The gods were then (mostly) banned from the picture, for explanations of natural phenomena were sought within nature itself, with physical reasoning replacing myth.

THE IONIANS

By the sixth century B.C. trade among various Greek states had grown in importance, and wealth led to the improvement of their cities and lifestyle. In those days the center of activity was Miletus, a city-state located in southern Ionia, now the Mediterranean coast of Turkey. It was here that the first school of pre-Socratic philosophy flourished, marking the starting point of the great intellectual adventure that led, two thousand years later, to the birth of modern science. According to Aristotle, Thales of Miletus was the founder of Western philosophy. The Alexandrian chronographer Apollodorus (second century B.C.) placed his birth in 624, while the great Greek historian Diogenes Laertius (third century A.D.) placed his death during the fifty-eighth Olympiad (548–545) at the age of seventy-eight. His reputation was legendary. He is said to have once used his astronomical and meteorological knowledge (most probably borrowed from the Babylonians) to forecast an excellent olive crop for the following year. Being a practical man, he then raised money during the winter to rent all the olive presses in the region. When summer arrived, the olive growers had to pay him for pressing their olives into oil, making Thales a fortune in the process.

He is also supposed to have predicted an eclipse of the Sun, on 28 May, 585 B.C., which ended a war between the Lydians and the Persians. When asked to define what was difficult in life, he answered, "To know thyself." When asked what was easy, he answered, "To give advice." No wonder he was known as one of the Seven Wise Men of Greece. He was not always practical, though. One day, while lost in abstract speculation, he fell into a well, failing to notice the hole in the ground. This accident apparently hurt the feelings of an attractive Thracian slave girl standing nearby, who remarked that Thales was not even able to see what was at his feet (i.e., her), because celestial matters were of more importance to him. The veracity of these and other stories surrounding Thales is much debated, as nothing written by him survived, which is true of many pre-Socratic philosophers. The evidence that is available to us comes secondhand from scarce fragments preserved by authors who wrote centuries after the death of these philosophers, from Plato in the fourth century B.C. to Simplicius in the sixth century A.D. A case in point is Aristotle's tendentious discussion of pre-Socratic ideas found in his *Metaphysics* or in *De Caelo*, "On the Heavens." While his writings must be used as a basic source of information, we should keep these limitations in mind as we explore the work of these philosophers.

The main concern of the Ionian philosophers was the composition of the cosmos. What is the basic stuff the Universe is made of? Thales's answer was that all was water. It is probable that apart from being influenced by Homer and by the great civilizations of the Near East, his choice of water as the basic substance was due to its inherent mutability; water cycles from sky to land to ocean to sky in a continuous flow, changing from liquid to vapor and back. As we and most living forms are nourished by water, so was the Universe, itself considered by Thales a living organism.

This organic view of the cosmos was quite remarkable in that it attempted to unify the way nature works and our own physiology. When Thales said that "all things are full of gods," or that magnetism can be attributed to the material having a "soul," he was arguing intuitively that the many natural phenomena are due to some inherent tendencies within the objects themselves. In fact, the word *soul* here must be understood as a sort of life principle by which *all* things are animated, and not in its modern religious sense. Simplistic as these ideas may seem to us, their importance cannot be overstated. Thales asked a new kind of question, and tried to find an answer without invoking mythical gods or other supernatural causes. In trying to explain the complicated workings of nature by searching for a unifying principle *within nature itself*, Thales set himself apart from the past, inaugurating the Western philosophical tradition.

Thales's immediate successor was Anaximander, also of Miletus, roughly fourteen years his junior. He carried Thales's ideas to a higher level of sophistication, postulating that the Universe was infinite in extension and duration, with a cylindrically shaped Earth placed at its center floating in air. He even went as far as claiming that the ratio of the diameter to the height of the cylinder was one-third. The Sun was a hole in a huge wheel surrounding the Earth with fire inside its rim. As the wheel turned so did the hole, explaining the motion of the Sun around the Earth. Eclipses were due to partial or total covering of the hole. The same basic explanation was used for the phases of the Moon, also a puncture in another cosmic wheel. Finally, the stars were tiny holes in a third cosmic wheel, which Anaximander curiously placed closer to the Earth than the Sun and Moon.

Bizarre as this imagery is, it represents the first attempt at a mechanical model of the Universe. In the words of Arthur Koestler, the driving mechanism was now viewed as "the wheels of a clockwork." According to Anaximander, the raw material of the Universe was not water or any other familiar substance but something intangible, the Boundless, "from which come into being all the heavens and the worlds in them." Note the use of plural; indeed, since the Universe was infinite in extension and duration, an infinite number of "worlds" had already existed before ours, only to be dissolved into the primary matter before others appeared. This dynamic picture of an infinitely old Universe where matter forever appears and disappears reminds us of the Hindu myth of the dancing Shiva. However, here there is no Creator, no God or gods responsible for the eternal cycle of creation and destruction. For Anaximander, the Universe danced alone.

Anaximander's disciple was called Anaximenes. In the true spirit of the Ionian school, Anaximenes also postulated the existence of one basic substance responsible for all change, while remaining unchanged itself. Challenging his masters, he believed that air was the raw material of the Universe. As its density changed, air composed different substances. Thus, when very rarefied it made fire, when thicker it became wind, then water, earth, and stones as its density increased. He is also credited with the idea that the stars were attached to a crystalline sphere that turned around the Earth. Being transparent, crystal spheres were a more plausible explanation for the heavenly motions than Anaximander's punctured wheels, which, of course, nobody could see. (Anaximander explained that they were perpetually surrounded by mist.) These crystal spheres would become a key component in astronomical models for another two thousand years.

The Milesians (this trio of philosophers from Miletus) were not the only ones asking questions about the Universe at this time. Besides other Greek thinkers that we will soon review, Siddhartha Gautama (the Buddha) in India was preaching on the ways to purge the self from greed in order to achieve Nirvana, while in China Lao Tse was transcending our polarized perception of reality in the mystical unity of Tao, and Confucius was establishing moral principles of government and life in society. The sixth century B.C. was a turning point in the history of civilization. A wind of awakening was blowing across the planet, calling the human mind to contemplate the inner workings of the soul and of the Universe.

The last Ionian thinker of importance to us was Heraclitus of Ephesus, who flourished around 500 B.C. Although Miletus was destroyed by the Persians in 494 B.C., the ideas of Thales and his followers reached Ephesus, located slightly to the north. Known from fragments of his teachings quoted by other authors, including Aristotle and Plato, Heraclitus wrote mostly in riddles that were notoriously hard to understand, and he became known as "the dark one." His style was often sarcastic and critical of other authors' ideas and intelligence. By the end of his life he had become a hermit, entirely withdrawn from the world. Falling ill of some edematous swelling, Heraclitus went to the village doctors to ask for help. Instead of simply stating his problem, he started to discourse in riddles that the doctors couldn't understand. Discouraged, Heraclitus buried himself in manure, hoping that the heat would evaporate the swelling. His treatment didn't work and he died, alone and dirty, at the age of sixty.

Although much of Heraclitus's thought is available to us only in fragments, his fame lies mostly in the teaching that "all things are in flux and nothing is stable," as Plato wrote in *Cratylus*. In his most famous utterance he stated that it is not possible to step in the same river twice. Extending this idea from nature to human life, he emphasized the importance of tension between opposites and their complementarity as the driving force behind all things, in such statements as these:

Beginning and end, on a circle's circumference, are common.
In us the same is living and dead, awake and asleep, young and

old, for these, transformed, are those, and those, transformed,
are these.

Men do not understand how, being pulled apart, it is in accord
with itself: a harmony, turning back on itself, as in the bow and
the lyre.

Balance is achieved by the necessary complementarity between
opposites, which he called *Logos*, as the bow that must be arched
back to propel the arrow forward. With some liberty, we could call
Heraclitus a Greek Taoist, although one must be careful interpreting these fragments out of context.

For Heraclitus the basic substance was fire, possibly due to its
power to transform things, to put them in motion. However, his
primary focus was on the change generated by the tension between
opposites, while for his fellow Ionians change was a secondary manifestation of the primal substance. In this Heraclitus disagreed with
the teachings of Thales and his followers. His universe was eternal,
uncreated, and in a perpetual state of flux: "This cosmos was made
by no god or man, but always was and is and will be: ever-living fire,
kindling in measures and quenching in measures." The objects in
the heavens were bowls containing fire, with the Sun the hottest and
brightest. As the bowl containing the Sun turned its back to us,
eclipses occurred. The phases of the Moon were due to the slow
turning of its containing bowl. Though it is not clear from the fragments if Heraclitus really meant these ideas as a true model of the
skies, his perception of nature as a dynamic, ever-changing entity
played an important role in later Greek thought.

THE ELEATICS

While Heraclitus was teaching that all was change, completely
opposite ideas were being developed in the southern Italian city of
Elea. Parmenides (*c.* 515–450 B.C.) stated instead that change is illusory, that what *is* cannot change, for if it does it becomes what it is
not. He dismissed the Milesian ideas rooted in the continuous transformation of the primal material into the myriad forms of matter
appearing in nature as relating to a superficial reality. The true
reality was static and immutable, the all-pervading, continuous Eon,

or Being. No holes or gaps could exist in the true nature of Being, since they would be equivalent to Non-Being. An image that comes to mind to represent this philosophy is that of the surface of a perfectly still lake stretching in all directions.

Whereas the Milesians searched for knowledge on the basis of empirical observations of natural phenomena, from the outside in, Parmenides's approach was from the inside out. In trying to understand the true essence of reality, he used abstract logical reasoning to arrive at the conclusion that the answer was not in everlasting change, but in the absence of change, the static plenitude of Being. Parmenides wrote that the absolute Being "neither was nor will be, because it is in its wholeness now, and only now." Thus, Eon could not have been created by something, for this assumes the existence of another Being, or even from nothing, as this would be Non-Being, which cannot exist. Eon just is.

But how did Parmenides and his followers, known as the Eleatics, reconcile their monistic doctrine of no change with the continual changes in nature? Actually, they didn't; instead they tried to prove that movement or change was an illusion of the senses. Perhaps the best illustrations of this effort are the paradoxes of Zeno, a disciple of Parmenides. His method is known in the scientific circles as "infinite regression." The reason for this name will become clear in a moment.

The most famous of Zeno's arguments against the reality of continuous motion involves Achilles (the hero from Homer's poem) and a turtle. Zeno postulated that if the turtle starts ahead of Achilles in a race, then he will never be able to reach and pass the turtle to win the race. Since it is motion that would lead Achilles to an eventual victory, if he can't win the race by moving faster than the turtle, motion doesn't exist. This is how his proof goes: By the time Achilles covers the initial distance between himself and the turtle, the turtle has moved farther away. As Achilles races to cover that new distance, the turtle moves farther away, and so on ad infinitum. It would take an infinitely long time for Achilles to reach the turtle!

This is a very disturbing conclusion that, at first, leaves us uneasy. How can reason be so obviously contradicted by the senses? The simplicity of Zeno's arguments must have given his adversaries many headaches. But fortunately, they are flawed: Even though mathematically we can divide the distance between Achilles and the turtle into smaller and smaller segments, in order to describe motion we should also divide time into smaller and smaller segments. It is the ratio between distance and time, the velocity, that is important here. And if you divide a small number by another small number, the result is not necessarily a smaller number. For example, $4/2 = 2$, but so does $2/1 = 2$, and $0.2/0.1 = 2$, and so on. Since Achilles's velocity is greater than the turtle's, he will cover a greater distance in the same amount of time, and beat the turtle. Motion is an illusion only in the abstract world of the Eleatics.

Nonetheless, modern physics and science in general owe a great deal to the Eleatics' principles. The task of physics is to find universal laws that describe natural phenomena we observe in everyday life and in the laboratory. By calling these laws universal, we are assuming that they are valid everywhere in the Universe and since the beginning of time. This assumption is rooted in our belief that nature is indeed immutable at some deeper level, that the laws we conceive to understand its behavior do not change.* Like Parmenides's Eon, they exist independent of all change they promote. This immutability of the fundamental laws of physics is what makes a rational approach to the study of nature possible. An Eleatic philosopher would probably say that by uncovering the laws of physics we are uncovering the essence of Being and would welcome modern scientists to walk inside the fortified walls of Elea and discuss the many facets of the One. I wonder if Zeno would agree to a race. . . .

THE PYTHAGOREANS

Pythagoras is perhaps the most legendary philosopher of antiquity. Shrouded by mystery, considered by some of his followers to be a demigod capable of superhuman feats, such as working miracles, talking to demons, and descending into the Underworld (and coming back to tell others about it), Pythagoras and his pupils forged a comprehensive synthesis of philosophy and religion, of the rational and the mystical, that is one of the greatest achievements of human

---

*Do we conceive or do we discover the laws of physics? Are they an invention of the human mind, or are they out there to be discovered by us and any other form of intelligence? We will get back to these questions later.

civilization. His religious philosophy has influenced some of the greatest philosophers and scientists of history, including Plato and Kepler. Some authors consider him the founder of science as we know it, or more broadly, the founder of Western European culture. As such, the Pythagorean legacy plays a very important role in the rest of this book.

Pythagoras was born between 585 and 565 B.C. on the island of Samos, located between Miletus and Ephesus. His father was a jewel engraver, from whom he must have learned the importance of geo-metrical forms and proportions and their association with symmetry and beauty. It is believed that he studied under Anaximander, and hence that he knew of the Ionians' search for a common substance underlying all that exists, and of the fiery-wheel model of the solar system with the cylindrical Earth in the center. He traveled ex-tensively in Greece, Asia, and Egypt, and must have learned of the various Eastern religions, and of the Babylonian mathematical tradi-tion. By 530 B.C. he had settled in the southern Italian colony of Croton, where he founded a religious brotherhood that quickly be-came a dominant spiritual and political force in the region. Because of its notoriously antidemocratic flavor, this local supremacy ended tragically about 495 B.C., when Pythagoras was forced to move to Metapontum and most of his followers were either exiled or killed. But by then "the word of the Master" had spread through several colonies around Croton, making its way into Athens by the fourth century B.C.

In order to understand the tremendous reputation surrounding Pythagoras, we have to examine his ideas free of modern notions that rule out the mixing of science with mysticism as utter nonsense. For the Pythagoreans there was no distinction between the two, one serving as a vehicle to the other. The basis for this union was their belief that "all is number," in a sense replacing the Ionian common primal substance with the search for numerical relationships be-tween all aspects of nature and of life. But contrary to the Ionians, this search was not merely rational but also mystical. If all things have form, and forms can be described by numbers, then number becomes the essence of knowledge, the gateway to superior wisdom. If knowledge is the pathway to apprehend the divine, numbers be-come the bridge between human reason and the divine Mind.

The aim of the Pythagoreans was to achieve catharsis, a purging of the soul, an intoxication of the spirit by the beauty of numbers, with their infinite power to transcend the limitations of everyday existence. In order to reach this stage, the members of the brotherhood had to follow strict ascetic rules of behavior, which included dietary and social restrictions such as not eating beans or meat, shunning butchers and hunters, and absolute loyalty and secrecy. As they progressed in their ascending path to ultimate knowledge, the disciples were initiated into the magical secrets of mathematics, in its pure abstract beauty.

Where does this curious and completely novel association of mathematics with the divine come from? One of the first discoveries of the Pythagoreans, usually attributed to Pythagoras himself, was that musical intervals can be expressed as numerical proportions. The basic intervals of Greek music can be expressed by simple ratios between the integer numbers 1, 2, 3, and 4. The tone of a plucked lyre string held at the middle is one octave higher than the tone of the whole string; at the two-thirds point it is a fifth higher; at the three-quarters point it is a fourth higher. Thus, numbers and simple relationships between them were behind the pleasing sounds that lifted people's spirits. Mathematics was thus associated with aesthetics, numbers with beauty.

This discovery is of tremendous historical importance. For the first time mathematics was used to describe a sensorial experience, as a vehicle to the workings of the human mind. From time immemorial music has been used as a medium to induce trancelike states in which the listener is lifted to a higher level of awareness. For the Pythagoreans the key to this phenomenon was the magic of numbers. Harmony was not simply due to pleasing sounds to the ear, but to numbers dancing according to mathematical relationships. Numbers also had form. For example, the number 4 was a square (think of the four vertices of a square), while the number 6 was a triangle (think of the three vertices of a triangle, and then add one point to the middle of each line joining the vertices). Adding square numbers produces square or rectangular numbers, as $4 + 4 = 8$, and the series of square or rectangular numbers is obtained by adding odd numbers in succession, $1 + 3 = 4 + 5 = 9 + 7 = 16 + 9 = 25$, and so on.

These relations led to the discovery of the famous Pythagorean theorem of right-angled triangles: The sum of the squared length of the two smaller sides is equal to the squared length of the longer side. (Curiously, Pythagoras himself was probably not involved with the proof of this theorem.)

The magic number for the Pythagoreans was 10, called the *tetraktys* (meaning roughly "fourness"), obtained by adding the first four numbers, $1 + 2 + 3 + 4 = 10$. That these numbers are also involved in the musical scales is no coincidence to the Pythagoreans. For it is the sacred number that describes true harmony. In another tremendous insight, the Pythagoreans extended this conception to the harmony of the heavenly spheres.* Surely, they argued, the Sun and the planets in their majestic beauty must satisfy the same harmonic rules that induce the musical communion with the divine. Thus, the distances between the planets would fall into the same ratios as the simple ratios of the musical scales. As the planets move along their orbits around the Earth-centered Universe, they would create music, a giant instrument playing God's melody, the harmony of the celestial spheres. Apparently only the Master himself could hear the celestial music. To the obvious criticism that this privilege provoked, the Pythagoreans answered, "What happens to men, then, is just what happens to coppersmiths, who are so accustomed to the noise of the smithy that it makes no difference to them."

The crucial point here is that the Pythagoreans started a new tradition in Western thought, the search for relationships between numbers in order to describe natural phenomena. This search is, in essence, what the physical sciences are all about. That they did this in order to ascend to a divine level of awareness is something that many later repudiated as empty mysticism. But there have been, throughout history, those who shared in this religious fascination with numbers and their magic power to inspire order in the apparently chaotic workings of nature, those I refer to as rational mystics in Chapter 1. The Pythagorean legacy inspired, directly or indirectly, some of the great giants who helped shape our modern view of the Universe. In understanding its historical importance, we

---

*It is generally believed that Pythagoras advanced the idea of the Earth's sphericity, although there is no agreement on this point among experts.

must differentiate between the individual motivations of scientists, which may have traces of Pythagoreanism, and the final results of their research.

The contributions of the Pythagoreans to astronomy did not stop at extending musical harmony to the heavenly spheres. Later Pythagoreans would boldly propose not only that the Earth moved, but also that it wasn't the center of the Universe. The first crucial step in this direction was taken by Philolaus of Croton, who around 450 B.C. narrowly escaped an attack by a pro-democracy mob that virtually ended the Pythagorean influence in southern Italy. He settled in the Greek town of Phlious, just west of Corinth, where he founded his own small Pythagorean circle. According to Philolaus, the Earth revolved around the "central fire," the "hearth of the Universe." This was supposed to be the source of all vigor and life, the altar from whence came even the Sun's heat. The Sun would merely distribute this heat to all the other luminaries. The central fire was invisible to us, since it was always opposite to the inhabited portion of the Earth, as shown on page 36. (Once we remember that we always see the same face of the Moon, this idea becomes a little less crazy.) Between the Earth and the central fire Philolaus proposed the existence of yet another celestial body, the *antichton*, or counter-Earth. This would also be inaccessible to the human eye, being always diametrically opposed to the inhabited parts of the Earth. After the Earth came the Moon, the Sun, the five known planets (Mercury, Venus, Mars, Jupiter, and Saturn), followed by the sphere carrying the fixed stars.

Philolaus may have had practical reasons for proposing this system. To an observer on Earth, the Sun and the planets seem to have two kinds of completely different motions. They turn daily around the Earth, like the stars; but contrary to the stars, which remain fixed in their relative positions, they have a motion of their own, turning at different rates around the zodiac, the narrow belt divided into the twelve constellations familiar from horoscopes. While the Sun takes approximately 365 days to complete one revolution, the periods for the planets vary from 88 days for Mercury to 29 years for Saturn. By making the Earth turn daily around the central fire, Philolaus could disentangle these two motions. As the Earth turned one way around the central fire, the whole sky would appear to turn

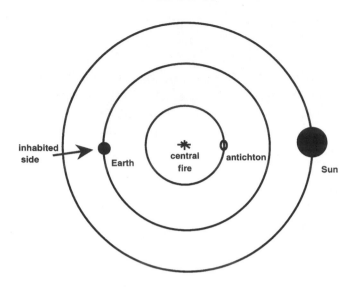

The system of Philolaus: night on Earth

the other way, like an observer on a merry-go-round who sees the whole fair turning backward. This would explain the daily rotation of the sky. Of course, the exact same effect is obtained by making the Earth spin around its axis, like a top. But this breakthrough would have to wait a little longer.

According to the historian of science Theodor Gomperz, "Nowhere else do we find a picture of the Universe at once so genial and sublime." Everything revolves around the central fire, the "citadel of Zeus," an expression of the Greek deep sense of symmetry and awe for a Universe ruled by divine power. There has been some debate as to why the counter-Earth has to be part of the model. Aristotle contemptuously remarked that Philolaus only invented it to bring the number of objects in the Universe to ten, the Pythagorean sacred number. Others have argued that the counter-Earth helped explain the large number of observed lunar eclipses, as it cast its shadow on the Moon's surface. Debates aside, the central fire of Philolaus was the first serious step toward proposing that the Sun is the center of the Universe.

THE ATOMISTS

If we now recapitulate the main ideas of the three pre-Socratic schools discussed so far, we see that we are set for a head-on collision. In one corner stand the Ionians, proposing that in its essence nature with all its manifestations can be reduced to a single material principle, be it water for Thales, the Boundless for Anaximander, or air for Anaximenes. Heraclitus then proposes that change is the true fundamental principle, being a consequence of the perpetual strife between opposites in their search for ultimate balance. For him, fire, this mediator of change, is the primal substance. In the other corner stand (as they can't move anyway) the Eleatics, arguing that change is illusory and that the true essence of all things is Eon, the all-pervading, static Being. Oblivious to the fight, the Pythagoreans dance to the divine harmony of numbers, immersed in their abstract mathematical mysticism.

It is clear that the pressing issue to be addressed by the philosophers of the mid fifth century B.C. is the problem of change. What is a young aspiring philosopher to do? Instead of picking sides and risking assault, perhaps the best way out is to try to reconcile these ideas into one all-encompassing scheme. This is precisely what Leucippus and Democritus, the founders of the atomistic school, chose to do, with brilliant results.

There is some confusion as to where and when Leucippus was born, and with whom he was associated. Probably he was from Miletus, like Thales, although some sources place his birth in Elea and others in Abdera, Thrace, the birthplace of Democritus (c. 460–370 B.C.), his most famous pupil. It is fairly certain that he was active between 450 and 420 B.C., and that he was Zeno's pupil. Note that these dates place Leucippus's prime and Democritus's birth after the birth of Socrates, making them the last of the pre-Socratic philosophers. Aristotle and his pupil Theophrastus credit Leucippus with the atomistic, or corpuscular, hypothesis, although his contributions are hard to distinguish from those of Democritus. In any case, Leucippus is usually credited with the main ideas behind atomism, and Democritus with their more detailed elaboration.

The great insight of the atomists was that change is not incompatible with the Eleatic ideal of an immutable Being. It is possible to describe the obvious changes observed in the world by assuming that the fundamental entity promoting those changes is itself unchangeable. This is where the idea of the atom, from the Greek *atomon*, meaning "that which cannot be cut," stems from. According to Leucippus and Democritus, the world is composed of an infinite number of indivisible, indestructible, and perfectly dense atoms, of an infinite number of shapes, that move aimlessly in the Void, or vacuum. As the atoms wander, they collide with each other, sometimes just recoiling from the collision and sometimes, if their shapes match, binding together to form all the complex forms of matter we observe. The atoms themselves are inert, passive Being, having no properties of their own. For example, the atoms of water and iron are in essence the same, differing only in shape; whereas the atoms forming water are smooth and round and thus cannot bind to each other, the atoms of iron are rough and uneven, explaining why they can conglomerate into solids. The importance of geometry in explaining the variety of forms in nature is a clear reference to the Pythagorean mathematical tradition. More important, by postulating the complementary existence of Being (the atoms) and Non-Being (the Void), the atomists achieved a beautiful synthesis of permanence and change, of being and becoming.

The atomistic hypothesis is, in a sense, the longest-lived of all pre-Socratic ideas. We all know from high school that the chemical elements are made of atoms, and that these atoms are themselves made of protons, neutrons, and electrons. Yet carrying the analogies with the ideas of Leucippus and Democritus to modern atomic theory too far can be misleading. True, the basic insight of the atomists, that matter is made of fundamental building blocks, is amazingly modern. However, modern atoms have very little to do with their pre-Socratic cousins. They are not infinite in number, and they are not indivisible. Atomic physics is an experimental science based on firm conceptual foundations, while the idea of experimental validation of physical concepts was completely foreign to the Greeks. Also, interpreting modern atoms as little billiard balls moving in empty space is essentially incorrect, as we will see later. If we extend the billiard ball analogy to the individual particles that make up the atom, the difficulties become even worse. The scientific importance of the Greek atomistic concept is mainly historical. It served as a source of inspiration to scientists trying to understand the structure of matter all the way into the early twentieth century. Nonetheless, with that proviso, there is indeed a path extending from Leucippus's insightful speculations to Rutherford's discovery of the nucleus and beyond.

Democritus proposed an interesting, though somewhat confusing, model describing the origin of worlds based on his atomistic hypothesis. First, atoms moved randomly in all directions. Due to this motion, collisions between atoms occurred, creating large vortices, or whirlpools, which were mainly composed of similar atoms. It seems that Democritus used circular motion as a kind of filter to group atoms of similar qualities together. As more and more like atoms conglomerated into a whirlpool, worlds were formed. Since there were infinitely many atoms, and as the Void itself was infinite, an infinite number of worlds existed at different levels of evolution and decay, ours being only one of them.

According to Diogenes Laertius, Democritus was one of the most prolific writers of antiquity. His works included not only the writings on physics and cosmology we have been discussing, but also zoology, botany, medicine, warfare, and ethics. He extended the atomistic idea to encompass the way we feel, see, taste, and even behave. An acid taste is composed of angular, small, thin atoms, while a sweet taste of round, larger ones. The sensation of white is caused by flat and smooth atoms that cast no shadow, while black is caused by rough and uneven atoms. Emotions are caused by atoms impinging on the atoms of the soul, and so on. There is a Grand Plan behind all this, which is ultimately to free humankind from the fear and superstition caused by the belief in gods and the supernatural. There is no purpose or design in nature, all basically boiling down to atoms randomly moving in the Void. Once we understand this simple fact, Democritus guarantees we will purge our soul of all heaviness, blissfully living in a state of "cheerfulness." For these ideas he was known as the "Laughing Philosopher." Nowhere is the social role of the atomistic hypothesis better expressed than in the wonderful poem *De rerum natura*, "The Nature of the Universe," by the Roman poet Lucretius (*c.* 96–55 B.C.):

This dread and darkness of the mind cannot be dispelled by the sunbeams, the shining shafts of the day, but only by an understanding of the outward form and inner workings of nature. In tackling this theme, our starting point will be this principle: *Nothing can ever be created by divine power out of nothing.* The reason why mortals are so gripped by fear is that they see all sorts of things happening on the Earth and in the sky with no discernible cause, and these they attribute to the will of a god. Accordingly, when we have seen that nothing can be created out of nothing, we shall then have a clearer picture of the path ahead, the problem of how things are created and occasioned without the aid of gods.

Or, a little later,

For the mind wants to discover by reasoning what exists in the infinity of space that lies out there, beyond the ramparts of this world—that region into which the intellect longs to peer and into which the free projection of the mind does actually extend its flight.

What lucid argumentation for the need of a scientific description of nature! There is complete faith that nature is comprehensible by reason alone, if we would only free ourselves from superstitions— the very attitude that makes science possible. In order for established (as opposed to speculative) science to be universal, it cannot depend on any particular system of belief, it cannot have room for subjective interpretation. The same equations have the same solutions for a Hindu, a Muslim, or a Jewish scientist. And this is certainly true inasmuch as the *practice* of science is concerned. It is, however, from this impersonal interpretation of science that much of its reputation as a purely rational activity comes; numbers are cold, equations are just collections of symbols describing some phenomenon to a handful of specialists who couldn't care less about the implications of what they do, and so on. But hidden in Lucretius's words we can also discern the other face of science, its human face. For the atomists, science is a response to a social need, a path to be followed by those who want to free themselves from the slavery of fear. Its power resides precisely in its universality, in its intrinsic a posteriori independence from subjective judgment.

IONIANS	**Thales** **(600)**	The first to look for explanations of natural phenomena within nature itself.
	Anaximander **(550)**	Proposed the first mechanical model of cosmos.
	Anaximenes **(520)**	Proposed the idea that objects in the heavens are carried around by invisible crystalline spheres.
	Heraclitus **(500)**	Viewed cosmos as being in a constant state of flux, a world of Becoming.
ELEATICS	**Parmenides** **(480)**	Believed in a cosmos of Being, immutable and permanent.
	Zeno **(460)**	Proposed several paradoxes suggesting the impossibility of motion.
PYTHAGOREANS	**Pythagoras** **(520)**	Developed the first ideas toward a mathematization of nature, based on his number mysticism.
	Philolaus **(450)**	Proposed the first model of cosmos with a "central fire," a precursor of a Sun-centered Universe.
ATOMISTS	**Leucippus** **(430)**	Believed that everything is made of indivisible atoms moving in the Void.
	Democritus **(400)**	Elaborated and further developed the atomistic hypothesis.

A time-table of pre-Socratic philosophers discussed in this chapter. Dates are approximate.

This does not mean that there is no room for individuality in science. Quite the contrary. I claim that it is with the *motivation* for doing science that the individual comes in; the longing for knowledge, the need to share in the excitement of discovery, of enlightening with the intellect the dark corridors of ignorance and fear, of lifting us above the obvious limitations of being human in this vast Universe. Springing from individual sources, science eventually achieves the universal. As we will see, this path is far from straight and narrow, far from being dispassionate. The true legacy of the Greeks is not only in their setting the conditions that made a scientific approach to nature possible, but to also, in retrospect, their showing the importance of individual beliefs in the scientific creative process.

PLATO AND ARISTOTLE

Around the same time that Democritus was describing the world in terms of indivisible atoms, Socrates was preaching that it was useless to try to understand the world before we understood ourselves. Instead of working as a stonecutter in his father's shop in Athens, Socrates would go to the market to preach to the young about the need for a new moral philosophy and for a reevaluation of the practices of government. In the words of Cicero, Socrates "called down philosophy from the skies." His influence was such that he was arrested by the nervous parents of the "corrupted young" and, being found guilty of impiety, was condemned to death by poisoning. This sad incident was a product of the confusing times in Athens by the late fifth century B.C.; in 404 B.C. the Peloponnesian War came to an end, and Athens finally surrendered to Sparta. Confronting great political turmoil, the people started to turn to more abstract values for comfort.

Born in 427 B.C., Plato very much embodied the spirit of the times. The fact that his uncle was one of the thirty tyrants who ruled Athens after its defeat served only to increase Plato's disgust for the current state of affairs. Being a pupil of Socrates, Plato believed that only by developing higher moral values firmly based on immutable truths could the malaise be improved. His self-imposed task was to formulate the new philosophy and then educate the future guardians of the ideal state, the philosopher-kings. Although he seems to have failed to enlighten any rulers, Plato's tremendous influence as a philosopher survives to this day. His academy, founded around 380 B.C., survived until A.D. 529 and can be thought of as one of the first universities in the world.

To Plato, the world is divided into two parts, the world of ideas and the world of the senses. The only true reality is in the world of ideas, composed of perfect and unchangeable Forms, since any concrete representation of an idea is necessarily inaccurate. For example, the drawing of a circle is never as accurate as the idea of a circle, which is perfect only in our minds. Thus a circle can only truly exist in the world of ideas. As a consequence of this doctrine, Plato had a certain disdain for observational science, as observations are always artificial.

This position may have cost Plato an exaggerated reputation as an enemy of science. Although he insisted on an abstract approach to reality, he also encouraged his pupils to study the heavens, even if only to help them find patterns that would guide them to more permanent truths. The importance Plato attributed to geometry was essentially Pythagorean in origin. When he said that "God ever geometrizes," he was translating the Pythagorean number mysticism into a geometrical mysticism, interpreting the existence of order in nature as the manifestation of superior design. This Cosmic Craftsman, or Demiurge, does not create the Universe or the matter in it (composed of combinations of air, earth, water, and fire), but uses its divine intelligence to impose the observed order in the world. The sensorial world is not as perfect as the abstract world of Forms, but it is the best possible world given that it reveals the operations of divine Reason. Thus, the study of astronomy is justified as a means of understanding the mind of the Demiurge, since the Universe reflects its intelligence. This teleology—the belief in the element of design in nature—is in stark contradiction to the atomistic doctrine of a random Universe, devoid of purpose.* Plato brings the need for a deity back into philosophy, although his Demiurge is a far cry from the anthropomorphic gods of Greek mythology. Possibly, the need for a God representing some intangible yet approachable ideal of order always increases in times of political chaos.

Plato's contributions to cosmology are somewhat hard to access due to the nebulous, mythical language of his book *Timaeus*. However, it is clear that he assumed that the heavenly bodies are spherical and that their motion is circular and uniform; that is, they rotate at constant angular velocity. According to Simplicius, he also posed the problem to his students of how to describe the observed irregularities and details of planetary motions in terms of combinations of simple circular motions. The search for a solution to this problem is

---

*I believe these two doctrines pretty much summarize the position of most people that think about those issues, including physicists, past and present.

what became known as *"saving the phenomena,"* that is, reducing the complicated motions of the heavenly bodies to simple circular motions, the inhabitants of the abstract world of the Forms. Thus, Plato's main contribution to cosmology was not perhaps as a developer of a specific system, but as an initiator of what would become the main concern of astronomy for centuries to come, the rational description of the heavenly motions.

It is clear from Plato's challenge to his students that he was well aware of the "irregularities" in planetary motion. Because this is a crucial factor in the development of astronomy all the way to Kepler in the seventeenth century, we should look briefly at what these irregularities are.

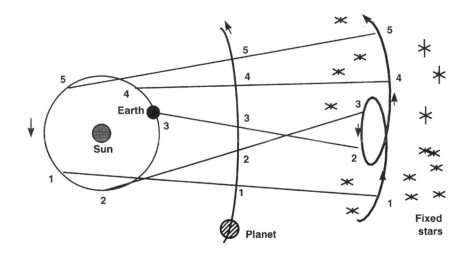

Retrograde motion: As the Earth orbits the Sun, the other planets will appear to trace a backward motion against the background of the fixed stars.

If we follow the path of a planet, say, Mars, for several months in the night sky, we observe that its motion is somewhat erratic. After progressing forward in its path with respect to the background of fixed stars, it appears to reverse its forward motion and go backward for a while before it moves forward again, as shown above. This backward motion, called *retrograde motion*, is really caused by the fact that the outer planets orbit the Sun at speeds slower than the Earth's. But for the Greeks, with their Earth-centered universe, the origin of retrograde motion becomes quite mysterious. The word *planet* comes from the Greek *planetes*, meaning "the wanderers."

Eudoxus (*c.* 408–356 B.C.), from the old Spartan city of Cuidus, proposed a brilliant solution to the challenge put forward by his master, Plato. Eudoxus's model shows not only advanced geometrical skills, but also an attention to observational detail previously foreign to the Greeks. His model was based on a series of concentric spheres centered on the Earth, itself a static sphere, in a sort of onionlike universe. Each of the five planets, Sun, and Moon were associated with a collection of imaginary spheres, four for each planet, and three for the Sun and Moon. Together with the outer sphere of the fixed stars, Eudoxus's model called for twenty-seven spheres in order to describe the motions of the heavenly bodies.

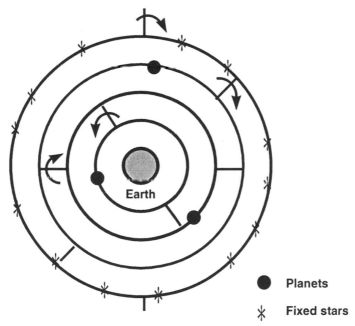

Earth

● Planets

⁎ Fixed stars

A simplified representation of the nested spheres of Eudoxus.

Briefly, this is how his system worked: Take a planet with its four spheres. Each one of the spheres is free to rotate about a different axis, with different speeds and directions (clockwise or counter-clockwise). The final motion of the planet from the point of view of an observer on Earth will be given by the combined motion of all the spheres. The outermost sphere accounted for the daily rotation of the whole sky by completing a turn westward in twenty-four hours. The next sphere accounted for the rotation of the planet about the zodiac; as different planets take different periods to complete a turn, this speed was different for each planet. Then came the two innermost spheres, which Eudoxus made to rotate with the same speed but in opposite directions about different axes.

The great discovery of Eudoxus was that the combined motion of the two inner spheres could approximately describe the peculiarities of retrograde motion. He showed that due to the combined rotation of the two inner spheres, the resulting planet's motion was like a figure-eight-shaped curve, called a hippopede. Adding the two outer spheres to this motion, Eudoxus achieved a pretty good qualitative description of the way the planets, Sun, and Moon moved from the point of view of an observer on Earth.

There are, of course, some obscure points in the model. There was no attempt to describe the nature of the spheres, or how motion was transmitted from one sphere to another. Another serious problem with the model, which was ultimately responsible for its downfall, is that it failed to explain why there is an apparent variation in the diameter of the Moon and in the brightness of the planets, both due to the fact that their distances from the Earth vary in the course of their orbits. However, the fact that Eudoxus managed to "save the phenomena" was a remarkable achievement, which pioneered a wholly new way of studying the skies. Before the model was abandoned in favor of more complicated models with epicycles, which we'll see soon, it was modified at least twice, first by a pupil of Eudoxus, Callippus of Cyzicus, and then by Aristotle himself. The modification by Callippus was to add seven more spheres in order to give a better description of retrograde motion. His model was in the same spirit as Eudoxus's in that the whole thing was an abstract mathematical construction. In other words, it was immaterial if the spheres really existed or not.

Not so for Aristotle. He wanted to convert the entire model into a mechanical one, with real spheres as opposed to abstract ones. His spheres would touch and transmit motion to each other, from outside in. In order to achieve this goal and still roughly account for the heavenly motions, Aristotle's model called for fifty-six spheres! But even with all this, the model did not attempt to explain the brightness variation of the planets, and was not considered very seriously, despite Aristotle's considerable reputation.

For over two thousand years, until the seventeenth century A.D., Aristotle's writings exerted a deep influence on Western thought. The history of science during this time can be summarized as a desperate attempt first to accommodate nature to the Aristotelian legacy, and then, during the last one hundred years of this long period, to supplant it. How can we understand the persistence of Aristotelian thought over such a long time span?

First, Aristotle's work had unprecedented breadth, covering topics from politics and ethics to physics, biology, and poetic theory. With his pupils, he not only compiled, classified, and built upon most of what was known by the fourth century B.C., but also helped develop whole new branches of knowledge, of which biology is the most notable example. A second reason is the commonsensical appeal of his ideas concerning the physical Universe. As opposed to Plato's abstract, mathematical Universe, Aristotle's Universe was concrete and physical. That his ideas were mostly wrong despite their apparent logic is mainly due to a combination of hasty generalizations and a lack of confrontation with observation. The final, and crucial, reason for Aristotle's grip on Western thought was the appropriation by the Christian Church of his ideas. Up to the twelfth century, Christian theology would be mainly influenced by Augustine's Neoplatonist thought, developed during the early fifth century in his books *Confessions* and *The City of God*. However, some elements of Aristotle's thought were also conveniently appropriated by theologians throughout this time. By the thirteenth century, Aristotle came back in full force, mostly due to the influence of Thomas Aquinas. His cosmology served quite well as a theology based on the separation of a decadent, ephemeral life on Earth, and a perfect, eternal existence in Heaven.

Born in 384 B.C. in Stagira, a city in Macedonia, Aristotle by the age of seventeen traveled south to join Plato's Academy, where he spent the next twenty years. Borrowing from Plato's teleological ideas, he looked for the final causes that explained not only the motions of the heavenly objects, but everything else that moved, from animals and plants to projectiles and people. All matter is composed of the four basic substances, earth, air, water, and fire, Aristotle surmised, to which he attributed four qualities, hot, cold, wet, and dry. Thus, water is cold and wet, while air is hot and wet, and so on. There were two kinds of motion, "natural" and "forced." A stone falls downward to return to its natural place, and if I want to throw it, I need to use a forceful motion. More important, natural motion is linear motion, such as the stone that falls straight down or the fire that goes straight up. This linearity of "natural" motion causes a serious difficulty in Aristotle's system, that is, how to explain the circular motion of the heavenly bodies. But this kind of problem would never stop a man like Aristotle; he simply postulates that the heavenly bodies are made of a fifth kind of matter, the *aether*, and that for this kind of matter "natural" motion is circular. This is uncreated, unchangeable, perfect matter, of a completely different nature from common earthly stuff, having no qualities like hot or wet. "Wait a minute," you say, "if they can't be hot, why do heavenly bodies shine?" "Because of the friction generated by their motion," Aristotle would quickly answer, with a hint of impatience.

By postulating the aether, Aristotle effectively divided his onion-like Universe into two realms, the sublunary (under the Moon) realm of natural linear motions and change, the realm of becoming; and the heavenly realm of unchangeable luminaries undergoing perfect circular motion throughout eternity, the realm of being. If you want to describe "motion without change," there is nothing more appropriate than circular motion, as it always returns to its starting point. Outside the sphere of the fixed stars is the sphere of the Unmoved Mover, the primary cause of all motion, the Being upon whom the whole Universe depends.

Aristotle's Universe is crucially different from other models we discussed before, such as the Pythagorean model with the central fire, or the infinite Universe of the Atomists. However, like the Atomists, he achieved a compromise between change and immutability; within the sublunary sphere the world is Ionian, with its emphasis on change and transformation. From there up it is Eleatic, for no change is allowed. His Universe is uncreated, eternal, and spatially finite. It is also a continuous Universe, without any empty space, or voids. God rules from the outside, with the static Earth being the farthest point from Him. This is the division of the Universe that the Catholic Church would later happily adopt. Unfortunately, the Church also adopted one of the worst trends in Platonic thought, its aversion to observational science, which remained practically nonexistent until the Renaissance.

THE SUN-CENTERED UNIVERSE OF ARISTARCHUS

A new era of astronomy was initiated with Eudoxus's model of nested spheres. Trying to satisfy Plato's request to "save the phenomena," attempts were made to explain the complicated motions of the heavenly bodies in terms of simple, circular motions. First, as in the models of Eudoxus and Aristotle, the attempts were mostly qualitative, since there were clearly conflicts between the predictions of the models and actual observations. But this situation would quickly change after Aristotle. The new models of the cosmos would attempt to *really* save the phenomena, that is, to get good quantitative agreement with observations. It didn't matter how complicated the models became, because they were just mathematical constructions devised to fit the data, not being attributed any physical reality. These attempts would eventually lead to the crowning achievement of Greek astronomy: Ptolemy's model of the cosmos, put forward during the second century A.D. It would dominate Western thought until the late sixteenth century practically unmodified and, besides a few developments in the Islamic world, mostly unchallenged.

The first step away from Eudoxus's nested spheres was taken by Heraclides of Pontus (*c.* 388–310 B.C.), a contemporary of Aristotle and probably, like Aristotle, a pupil of Plato. Heraclides is associated with two important astronomical discoveries. First he proposed (or at least was the first to clearly use) the rotation of the Earth around its axis in order to understand the apparent daily rotation of the skies. He made the Earth move again! I say "again" because we have encountered another model with a moving Earth, the central fire model of Philolaus, the Pythagorean. Both ideas were "easily" dismissed by Aristotelians, who argued that if the Earth moved, we would obviously notice it in the motions of falling objects or clouds left trailing behind. If I throw a stone straight upward, it will fall right back on my head. Clearly, argued the Aristotelians, while the stone goes up and down, the rotation of the Earth would carry me away from my starting point and the stone would miss my head. So much for the rotation of the Earth.

The second discovery attributed to Heraclides is that the inner planets, Mercury and Venus, orbit the Sun and not the Earth. Ironically, this discovery opened the path to two diametrically opposite developments in astronomy: the heliocentric (Sun-centered) model of Aristarchus and the geocentric (Earth-centered) model of Ptolemy. Heraclides probably proposed this modification based on the twin observations that the orbital periods of the inner planets are under one year, and that they are always "close" to the Sun in the sky. It is as if the Sun carries the two planets with it in its yearly trip around the zodiac. (Allusions to this theory, in typically ambiguous fashion, could already be found in Plato's *Timaeus*.) Although a clear step in the right direction, this idea was also dismissed by Aristotelians. Shifting the center of the orbits of Mercury and Venus to the Sun was a serious disruption of the Aristotelian order of the cosmos. The Earth, and only the Earth, could be at the center, the last lowly link in a continuous chain forged all the way up to the sphere of the Unmoved Mover.

Heraclides's proposal of the Sun as the orbital center for the inner planets may have inspired Aristarchus to complete the job and put the Sun at the center of *all* planetary orbits, including Earth's. In one of the most curious episodes in the history of ancient Greek astronomy, a Sun-centered Universe was proposed during the third century B.C., only to be forgotten for almost two thousand years.

Aristarchus was born in Samos, the birthplace of Pythagoras, around 310 B.C., the year Heraclides died. He was an extremely gifted mathematician, a meticulous observer, and clearly a courageous free-thinker, three rarely combined characteristics of great scientists. Of Aristarchus's works only one survived, *On the Sizes and Distances of the Sun and Moon,* in which he used clever geometrical arguments combined with astronomical observations to obtain the relative sizes and distances of the Sun and Moon. He showed:

1. The distance from the Earth to the Sun is about 19 times the distance from the Earth to the Moon

2. The diameter of the Sun is about 6.8 times larger than the diameter of the Earth

3. The diameter of the Moon is about 0.36 times the diameter of the Earth.

In fact, the correct numbers are for (1) 388, (2) 109, and (3) 0.27. Aristarchus was way off the mark on (1) and (2) not because of mathematical errors, but because of imprecise data, an obvious problem of naked-eye observations. In any case, the fact that he found the Sun to be considerably larger than the Earth possibly influenced his later decision to put it at the center of the cosmos.

We know of Aristarchus's proposal of a heliocentric Universe from the writings of Archimedes, arguably the greatest mathematician and inventor of antiquity, famous for running naked down the streets of Syracuse screaming "Eureka" after figuring out why things float.* In his short work *The Sand Reckoner,* Archimedes calculated how many grains of sand would fill the Universe, a very large number indeed. Because he needed a measurement of the size of the spherical Universe, he used Aristarchus's figures, the biggest Universe available in those days. His answer was that it would take 10^{63} (that is, a one followed by sixty-three zeroes!) grains of sand. Archimedes writes:

Aristarchus of Samos brought out a book consisting of some hypotheses, in which the premises lead to the result that the Universe is many times greater than now so called. His hypotheses are that the fixed stars and the Sun remain unmoved, that the Earth revolves about the Sun in the circumference of a circle, the Sun lying in the middle of the orbit. . . .

*In Archimedes's own words, "Any solid lighter than a fluid will, if placed in the fluid, be so far immersed that the weight of the solid will be equal to the weight of the fluid displaced." He used this idea to check if a crown ordered by his friend King Hieron II of Syracuse as a present to the gods was indeed made of solid gold as the King wished, or if, instead, the manufacturer cheated by mixing silver with the gold, as some suspected. Archimedes clearly showed that the manufacturer cheated.

It is now known that Copernicus, the man who brought the Sun back to the center of the Universe in the sixteenth century, was familiar with Aristarchus's work. The obvious question is, why was Aristarchus's heliocentric model forgotten for so long? One explanation, of a technical nature, is that if the Sun were the center of the Universe, an effect called *stellar parallax* should be detected, and it wasn't.

We can understand what stellar parallax is by studying the diagram below. Consider an observer on Earth measuring the position of a relatively near star with respect to some more distant background constellation. She will notice that the star seems to move with respect to the background constellation at different times of the year, an obvious consequence of the fact that we orbit the sun. The problem is that the stars are so far away, that the angular variation in the position of the nearest star is tiny, impossible to measure with the naked eye. Stellar parallax, the definitive proof that we orbit the Sun, would be detected only in 1838 by Friedrich Bessel. Had it been detected by the Greeks, possibly the entire history of astronomy and science would have been different.

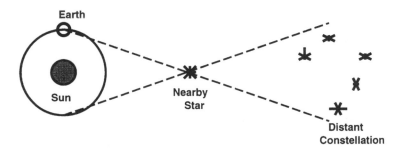

Because of stellar parallax, the position of a nearby star appears to move with respect to distant constellations as the observer's position changes with the Earth's orbit.

This issue aside, the most probable reason for the neglect of Aristarchus's model was again the powerful grip of Aristotelian physics. For an Aristotelian, putting the Sun in the center of the cosmos was an obvious mistake, given that the Sun, like the other heavenly bodies, is made of aether, which pursues a circular motion and thus cannot possibly be at the center. How would you then explain why things fall to Earth? And how could the Earth have the same "orbital status" of the planets, being made of something else? With these arguments, the doors closed on the Sun-centered Universe.

WHEELS WITHIN WHEELS: THE PTOLEMAIC UNIVERSE

The next great advance of Greek astronomy came with the advent of epicycles. These were presumably invented by Apollonius of Perga (265–190 B.C.), a mathematician compared only to Archimedes in his brilliance. An epicycle can be best visualized if we think of a Ferris wheel designed by a fiendish engineer; the chairs attached to the big wheel are allowed to rotate completely, the head of the poor passenger describing a complete circle. The circle is called the *epicycle*, while the big wheel is called the *deferent*. Now imagine that the evil engineer (no physicist would be this perverse) switches the power on, and the big wheel starts to turn. The head of the passenger would describe a spiral curve about the big wheel, as shown in the diagram on the left.

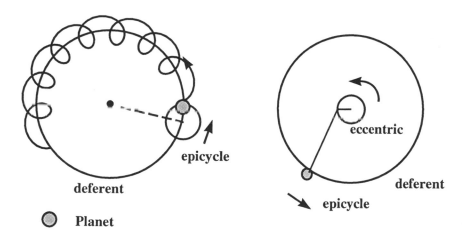

Comparison between epicyclic and eccentric motion. Both models give similar observational results.

Now substitute for the center of the big wheel the Earth, and for the passenger's head a planet. From the point of view of the Earth, the planet clearly displays retrograde motion. Its distance to Earth also varies, "explaining" the change in brightness. By combining the two spinning wheels, Apollonius described the observed peculiarities of planetary motion. He also proved that a similar spiral motion can be obtained if the "chair" is kept fixed and, instead, the center of the big wheel is made to wobble about a small circle, as shown in the diagram on the right. This off-centered motion is known as *eccentric* motion.

However remarkable his constructs, Apollonius was a true "theorist" and does not appear to have tried to apply his ideas of epicycles and eccentrics to describe the observed motions of heavenly bodies. It was Hipparchus, the greatest astronomer of antiquity, who applied epicycles to describe the motions of the Sun and Moon around the Earth. Born in Nicaea (today's Iznik, in Turkey), Hipparchus was active between 150 and 125 B.C. By this time the Romans had conquered all of Greece, and the center of intellectual activity had shifted from Athens to Alexandria in Egypt, founded by Alexander the Great two centuries earlier.

Let's backtrack to briefly retrace the expansion of Greece toward the East. Thanks to the military genius of Alexander's father, Philip, inventor of the cavalry as a striking force and of the tightly packed infantry formation known as the "Macedonian phalanx," Alexander had extended the boundaries of the Greek empire all the way through Egypt, Babylon, and into India, disseminating Greek culture over this vast domain. After Alexander's early death at thirty-three in 323 B.C., his empire disintegrated into fiefdoms controlled by his many generals.

Luckily, Alexandria was taken by the Macedonian general Ptolemy I, a close friend of Alexander's and a follower of Alexander's tutor, none other than Aristotle. Ptolemy took the Egyptian title of Pharaoh, even though his court was entirely Greek. He founded the first permanent endowment for the sciences, the Museum of Alexandria. Aristarchus, Apollonius, Archimedes, and Hipparchus were frequent visitors, as were the great geometers Euclid and Eratosthenes, who first measured the diameter of the Earth to within fifty miles of its true value, and Hero, the inventor of the first steam engine. Alexandria as a center of research and knowledge would remain dominant for several centuries, dying out by A.D. 200.

Hipparchus did much more for astronomy than pioneering the use of epicycles. Among other accomplishments, he created that much-beloved topic of high school math, trigonometry; obtained the best astronomical data of his time by combining old Babylonian observations with his own; invented the astrolabe, an instrument used to measure the positions of objects in the sky; and discovered the precession of the equinoxes—that is, the fact that the Earth wobbles about its axis of rotation, like an unbalanced spinning top. Somehow, he didn't attempt to apply epicycles to describe planetary motion, despite being critical of previous models based on nested spheres for their incompatibility with observational data. This was to be the role of Claudius Ptolemy, the last great Alexandrian astronomer (not to be confused with the general), who lived three hundred years after Hipparchus.

Not much is known about Ptolemy's life, although we do know that he flourished between A.D. 127 and 141 in Alexandria, by then a Roman province. His great work, called by later Arab astronomers *The Almagest*, "The Greatest," became the standard astronomy text until the late sixteenth century. In it Ptolemy builds upon Aristotle's ideas and Hipparchus's astronomy to achieve a comprehensive description of the motion of *all* heavenly bodies, which was in good agreement with observations. It is the crowning act of Plato's appeal for saving the phenomena, a tremendously complicated Universe built upon variations on the epicyclic idea, a giant machinery of wheels within wheels, eternally turning under the guidance of the Unmoved Mover.

What motivated Ptolemy to take up Plato's challenge of saving the phenomena so many years after his predecessors? For him, as for Plato and Aristotle centuries before, the heavenly bodies were divine. Furthermore, the order perceived in the Universe was a manifestation of superior Reason. The study of the heavens would lift the astronomer from the crudeness of everyday life into the realm of the gods, pointing the way to a higher moral and ethical existence. In a study deeply related to moral philosophy, by investigating the workings of the cosmos, the astronomer would be in touch with the divine.

Apart from devising a method to predict the positions of the planets, Ptolemy further modified the epicycles of Hipparchus to include yet another point, the *equant*, about which the invisible arm responsible for moving the big wheel (the deferent) was made to turn. The center of the deferent would be midway between the Earth and the equant, as shown in the diagram below. The center of the epicycle travels with uniform speed about the equant, and not about the center of the deferent or the Earth, as in Hipparchus's scheme. By adjusting the distance between the center of the wheel and the equant for different planets, Ptolemy was able to reproduce a host of irregularities presented by planetary motion. But at a cost. His model violated one of the Platonic dogmas, that the motion of the heavenly bodies should be uniform about the Earth. That this didn't bother Ptolemy is proof that his main concern was with saving the phenomena, and that he was prepared to be much more cavalier with his physics than with his astronomy.

Ptolemy's model was a tremendous success. His whirling wheels could not only describe the motion of the Sun, Moon, and planets, but also predict with reasonable success their future location, to the delight of astronomers and astrologers alike. However his model was partially forgotten in the Western world during the next eight hundred years or so, until the Arabs brought it back to Europe at about A.D. 900. It would then dominate astronomy (and astrology) until the sixteenth century, when Copernicus proposed his heliocentric model.

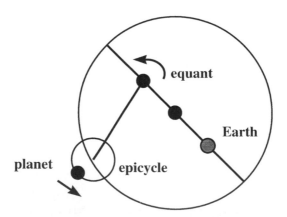

Ptolemy refined the use of epicycles by introducing the equant. His epicycle revolved around the equant, rather than the Earth.

Before we leave Ptolemy and the Greeks, I should say a few words about the crucial role astrology played in the initial development of astronomy. As early as 2000 B.C., the Babylonians clearly attributed a magical reality to the influence of the Moon, Sun, and planets (especially Venus) on public affairs and the lives of their rulers. This influence was brought to the level of the common man by the Greeks, who, in a sense, personalized astrology, blending it with their own mythology, in which heavenly bodies were represented by different gods. An astrologer was an interpreter of the motion of the gods, a bridge between the divine and the human. It was clearly a position of power and prestige.

For effective divination, accurate prediction of the positions of the planets, Sun, and Moon with respect to the twelve constellations of the zodiac was crucial. Thus, to the Platonic appeal for saving the phenomena we must add the astrological need for a predictive astronomy. Ptolemy wrote a complete work on astrology, *The Tetrabiblos*, in which he cleverly blended the uses of astronomy and astrology as twin paths in the search for a deeper state of tranquility, attainable only from superior knowledge. For Ptolemy, while astronomy had a moral value, astrology, with its forecasting powers, "calms the soul by experience of distant events as though they were present, and prepares it to greet with calm and steadiness whatever comes."

Astrology continued to be important in later Roman society, before its repression by Christianity started during the fourth century A.D., most notably voiced in Augustine's work *The City of God.* At the center of this later debate was the question of free will. In astrology the Universe is essentially mechanistic, and there is no room for individual choice, since the future is decided by the motions in the heavens. This clashed, of course, with the omnipotence of the Christian God, making astrology untenable. Attempts to solve the impasse stressed that "the stars don't impose, only suggest," leaving to the individual the final choice of the course taken, guided by God through prayer.

Despite similar repression by their theologians, astrology was also practiced in the Islamic world; and with the Muslim conquest of Sicily and Spain, astrology would reenter Western Europe, undergoing a revival during the twelfth and thirteenth centuries. It would be part of the standard curriculum of the first medieval universities

in Bologna, Paris, and Oxford, and would be an early source of in-
come and inspiration (as well as conflict) to Johannes Kepler, who
in the early seventeenth century devised the first mathematical laws
of planetary motion. With the formulation of Newtonian mechanics, astrology and astronomy would be permanently divorced, at
least in the minds of scientists. However, astrology remains as fascinating to us as it was to the Greeks, as can be seen by its tremendous
popularity. For those who seek astrology for comfort and as a vehicle of self-discovery, Ptolemy's words are as valid today as they were
eighteen centuries ago.

Plato **(380)**	Proposed that the true nature of reality lies in the world of ideas. Challenged pupils to "save the phenomena."
Eudoxus **(370)**	Proposed an Earth-centered model of nested spheres.
Aristotle **(340)**	Suggested "common-sense" explanations of natural pehnomena. Divided cosmos in two realms.
Heraclides **(350)**	First to suggest the Earth's rotation. Also proposed that Mercury and Venus orbit the Sun not the Earth.
Aristarchus **(280)**	First model of a Sun-centered Universe. Also computed the distances to Sun and Moon.
Archimedes **(250)**	One of the greatest mathematicians and inventors of antiquity.
Apollonius **(230)**	Invented the epicycles, although never applied them to astronomical problems.
Hipparchus **(140)**	Applied epicycles to explain the motions of the Sun and the Moon. Considered the greatest astronomer of antiquity.
Ptolemy **(130 A.D.)**	Developed a complete model of the cosmos using epicycles.

A time-table of Greek philosophers from Plato to Ptolemy
discussed in this chapter. Dates are approximate.

This review of some of the most important Greek ideas is by necessity brief. It would be impossible to do true justice to the ancient
Greeks, for their intellectual and artistic legacy fills many volumes.
But I hope to have given you an idea of their tremendous creativity.
Perhaps more than anything else, the Greeks taught us the importance of asking questions. Their love for reason and their belief in
its power to confront the mysterious is at the very foundation of a
scientific approach to nature. Humankind is not to shy away from
knowledge. Fear must be answered with reason, and not more fear.
This is the key to wisdom.

As we move on to the early Middle Ages, we will see that this curiosity about the world all but disappears. The ascendancy of the
Church and the decline of Rome redirected the worries of "educated" people toward abstract theological questions; the seeds
planted by the Greeks were to lie dormant for quite a while. Some
scientific progress was made during this time, however. The Arabs,
in particular, improved on Ptolemy's model considerably, and
raised mathematics to new levels of sophistication. But the Universe
remained essentially Aristotelian, Earth-centered, finite, and divided between the two realms of being and becoming.

The only acceptable wisdom was of a theological nature, and any
questioning about the workings of the world was considered superfluous and dangerous to the salvation of one's soul. The state of astronomy was so regressive that for seven hundred years, from roughly
A.D. 300 to 1000, the Earth was once again considered to be flat! But
as the Muslims brought back to Europe the texts of Aristotle, Euclid,
Archimedes, Ptolemy, and many others, another wind of awakening
started to blow, slowly freeing the mind from the slumber of the
Dark Ages. It would blow softly from the late Middle Ages (thirteenth century) to the early Renaissance (fifteenth century), gradually changing into a true hurricane by the early seventeenth century
in the hands of Giordano Bruno, Galileo, Kepler, and William Gilbert.
Civilization witnessed a true rebirth induced by the rediscovery of
Greek wisdom. As Aristotle wrote prophetically, "We cannot help
believing that the same ideas recur to men not once nor twice but
over and over again."

PART II

○

THE AWAKENING

3

THE SUN, THE CHURCH, AND THE NEW ASTRONOMY

What? Can I not everywhere behold the Sun and the stars?
Can I not under any sky meditate on the most precious truths?
— DANTE ALIGHIERI

The awakening was slow, a lazy spring fighting the cold embrace of winter. For centuries the medieval mind was immersed in dogmatic theology. The wisdom of the past was forgotten, the splendor of the Greek and Roman civilizations a distant memory, condemned by the Church as pagan knowledge, the root of all evil. During the fifth century, St. Augustine forged a tenuous link to the past through false Neoplatonism that despised any interest in natural phenomena, while encouraging a purely rational approach to religious themes. All answers to cosmological or astronomical questions were to be found in the Bible. The heavens were not spherical but shaped like a rectangular tent (a tabernacle), because in Isaiah it states that God "stretches out the heavens as a curtain and spreadeth them out as a tent to dwell in." Similarly, the Earth was viewed as either a rectangle or a disk, depending on what biblical source was consulted.

Why did this happen? What is the relation between the ascendancy of the Church and its almost complete severance with the wisdom of the classical world? To answer this question, we have to consider the political situation in Europe around the time of St. Augustine.

By the fourth century A.D., the once mighty Roman empire was crumbling, both from within and from without. Divided into a Latin-speaking Western empire and a Greek-speaking Eastern empire (known as the Byzantine empire) roughly where the Danube River meets Serbia and Romania, the Romans suffered continuous waves of attack from several Germanic tribes, such as the Vandals and the Goths, and the Persians in the east. From the inside, corruption and moral decadence combined to undermine the "Roman pride." Change was badly needed, something that would restore the sense of direction to a confused and divided society.

In 324 Constantine the Great, ruler of the Eastern empire, embraced Christianity as his faith. He changed the name of his capital from Byzantium to Constantinople (now Istanbul, Turkey), which quickly became the new center of Christian power. As the Byzantine empire grew in strength, it attempted to recapture the West from the Germanic invaders, spreading Christianity to new tribes and strengthening the many already existing Christian communities throughout Europe. Even though the empire eventually failed and Rome fell to the Germans during the fifth century, Christianity survived, guided by such leaders as St. Augustine and Pope Gregory I (590–604). It became the most civilizing influence in Western Europe, imposing religious discipline as an antidote to "barbarian pagan rituals." To lives filled with violence, pestilence, and unrest, the Church offered eternal salvation in Paradise. Its power was such that during the fifth century, when Attila the Hun wanted to march into Rome, the Patriarch persuaded him to turn back, something that no army in the world was powerful enough to do. In a sense, "the Church had conquered its conquerors."

The Church's condemnation of knowledge for knowledge's sake was based on the grounds of spiritual purity. The barbarism corrupting the body was the same that corrupted the mind; anything to do with the appropriation of information through the use of the senses was a sure route to depravity. The temptations of the flesh and all sensorial experience were dangerous distractions from the true path toward eternal salvation. Since an interest in the workings of nature linked people to external reality, it was deemed "pagan" knowledge, a corrupting influence on Christian virtue. In the words of St. Augustine,

At this point I mention another form of temptation more various and dangerous. For over and above that lust of the flesh which lies in the delight of all our senses and pleasures . . . there can also be in the mind itself, through those same bodily senses, a certain vain desire and curiosity, not of taking delights in the body, but of making experiments with the body's aid, and cloaked under the name of learning and knowledge. . . . Thus men proceed to investigate the phenomena of nature—the part of nature external to us—though the knowledge is of no value to them: for they wish to know simply for the sake of knowing. Certainly the theaters no longer attract me, nor do I care to know the course of the stars.

THE SLOW LIFTING OF THE VEILS

For seven long centuries, most of Europe was consumed by continual fighting among the many feudal lords. Political power was mostly decentralized, with the exception during the ninth century of the reign of Charlemagne, who was crowned by Pope Leo III as Emperor of the "Holy Roman Empire." Even though long gone, the grandiosity of Rome still lingered as the utmost symbol of power.

While Europe was lost in complete political disorder, the eighth century witnessed the flourishing of a new empire. Its borders stretched from North Africa and Spain in the west, through Egypt and Persia into Central Asia. In the Muslim empire, once again Aristotle and Ptolemy were read and discussed, and arts and architecture were praised and supported by the caliphs. The Arabs brought back to their domains a love for learning that had been long forgotten. Together with Jewish scholars, they forged a new cultural class in Spain that, in the next five centuries, was to change completely the intellectual landscape of Europe. Their enthusiasm toward the Greeks slowly diffused through Europe (it was thick, that medieval fog!), creating the conditions that later would blossom into the Renaissance.

During the twelfth and thirteenth centuries, while crusaders were trying to recapture the Holy Land from the Muslim East and magnificent Gothic cathedrals were being built in France, Aristotle and Ptolemy were captivating more and more minds. By the late thirteenth century several universities had been founded, and the need for better introductory texts on mathematics, philosophy, and astronomy grew. John of Sacrobosco's *On the Sphere* became a widely read book on elementary astronomy, as well as Latin translations of Arab texts summarizing Ptolemy's *Almagest*. Aristotle grew in prominence and, slowly, a renewed curiosity for the workings of nature began to emerge.

Largely due to St. Thomas Aquinas (1225–1274), Christian theology embraced Aristotelian ideas, and a new "Christian cosmology" was born. The Earth became spherical again, occupying the seat at the center of the Universe. There followed the nested spheres, connecting the Earth to God. The eighth sphere, that of the fixed stars, was followed by the *Primum Mobile*, the outmost mobile sphere. The tenth and last sphere was immobile, the *Empyrean sphere*, "the dwelling place of God and the elect." Since Lucifer sat on his throne in Hell, closer to people on Earth than God in Heaven, the medieval Universe was, in effect, "diabolocentric." Perhaps the best description of the late medieval cosmos can be found in Dante's *Divine Comedy* (1321), in which the poet retells his journey through the three realms of the afterlife: Hell, Purgatory, and Heaven. As he travels from Hell in the center to Heaven above, he crosses all the heavenly orbs in the order prescribed by Aristotle.

Sadly, even though Aristotle was rediscovered, his ideas were taken dogmatically, and any attempt to refute them called pointless. One of the chief occupations of late medieval theologians was to develop arguments to reconcile Aristotelian ideas with Christian dogma. After all, the Aristotelian cosmos was uncreated and eternal, while for Christians, God created the Universe, and earthly life would come to an end on Judgment Day. The usual strategy was to cleverly reinterpret Aristotle to serve the purposes of the Church; having created the intellectual climate that could promote the development of free thinking, the medieval scholars and churchmen quickly made sure that no changes were to be contemplated.

This was a source of despair to some free-spirited thinkers who refused to blindly accept the Aristotelian ideas. Roger Bacon (*c.* 1219–1292), an Oxford Franciscan, wrote, "If I had my way, I should burn all the books of Aristotle, for the study of them can only lead to a loss of time, produce error, and increase ignorance." He urged elsewhere that philosophers "cease to be ruled by dogmas and authorities; *look at the world!*" In his books Bacon speculated that powered machines would exist in the future, not only for transportation on land and sea, but also for flying. He emphasized the importance of mathematics and experimentation as a way to approach God's creation, making him one of the important influences in the early development of modern science. When I think of Roger Bacon, the image of a "science prophet" comes to my mind, a lone visionary announcing the impending doom of the medieval Universe.

Another thinker whose ideas were ahead of his time was Nicholas of Cusa (*c.* 1401–1464), made Bishop of Brixen (Bressanone, Italy) in 1450, and also a papal legate in Germany. In his famous work *De docta ignorantia,* "On Learned Ignorance," he claimed that wisdom lies in the impossibility of the finite human mind to understand the infinity of God, in whom all opposites are combined.* In order to transcend this limitation, he made extensive use of the "principle of the coincidence of opposites," claiming that apparent contradictions are unified at infinity, i.e., God. This had interesting cosmological consequences, for Cusa claimed that there cannot be a physical center in the Universe, since it is impossible to find the perfect center. This sounds very much like Plato's claim we encountered before, that a circle is perfect only as a thought construction. He thus removed the Earth or, for that matter, any other celestial body from the center. For Cusa, the center was the abstract seat of perfection. As such it had to be equated with God, as only in God could perfection be found. And since the opposites were united at infinity, God was also the boundary, or circumference of the Universe. Cusa's universe was delimited by theological arguments:

> Since the center is a point equidistant from the circumference, and since it is impossible to have a sphere or circle so perfect that a more perfect one could not be given, it clearly follows that a center could always be found that is truer and more exact than any given center. Only in God are we able to find a center which is with perfect precision equidistant from all points, for He alone is infinite equality.

*Recall our discussion in Chapter 1 about our limited perception of reality based on the distinction of opposites.

Although an inspiration to future generations of thinkers, Cusa's ideas were firmly planted in the past; Plato's Demiurge has merely been replaced by the Christian God.

Both Bacon and Cusa encountered trouble with their peers and superiors for daring to challenge the accepted cosmological framework; the huge and stable edifice of the medieval cosmos started to develop cracks in its very foundation. Suspected of promoting "dangerous novelties," Bacon was imprisoned between 1277 and 1279 by order of the minister-general of the Franciscans. Cusa's political rivals charged him with pantheism, and he was forced to write his *Apologia doctae ignorantiae*, "Apology on the Learned Ignorance," in 1449, in which he quoted Church fathers and Neoplatonists to justify his ideas.

When threatened, quoting ancient authority was always a safe way out. But not for much longer. A distinguishing feature of the forthcoming "Copernican Revolution," unwittingly initiated by Copernicus and carried through in earnest by Kepler and Galileo, was a complete change of attitude with respect to ancient authority. A new approach to the study of nature was about to be born and was to cause perhaps the most profound transformation in the human spirit since the sixth century B.C. This bridge between the old and the modern, forged through much courage, passion for truth, and brilliance, is the subject of the remainder of this chapter.

THE RELUCTANT REBEL

Some people become heroes against their will. Even though they may have ideas that are revolutionary, they may not know it, and if they know it, they may not completely believe them. Torn between exposing themselves to outside opinion or keeping a defensive low profile, they opt for the latter. The world is full of poems and theories hidden in the attic.

Copernicus is the most famous of these reluctant heroes in the history of science. The man who put the Sun back at the center of the Universe didn't want many people to know about it, possibly wary of criticism and religious persecution. He also put the Sun back at the center of the Universe for the wrong reasons. Unhappy with Ptolemy's failure to adhere to the Platonic dogma of uniform circular motion for the heavenly bodies, Copernicus proposed to abandon the equant and instead to put the Sun in the center of all orbits.

In trying to make the Universe adjust to Plato's requirement, Copernicus went all the way back to the Pythagoreans, resuscitating their doctrine of the central fire that culminated in Aristarchus's heliocentric universe. His thought reflects a willingness to shake the very foundations of the cosmological ideas of his time, but only in order to reach farther back into the past; he was, in short, a conservative revolutionary. He could never have guessed that, by looking so far back, he would be propelling civilization into the future. Had he lived to see it, Copernicus would have hated the revolution he unknowingly started.

Nicolaus Copernicus was born February 19, 1473, in Torun, Poland, the son of a wealthy merchant. When he was ten his father died, and he was taken under the custody of his maternal uncle, the powerful Lucas Waczenrode, the future Bishop of Ermeland. In 1491, a year before Columbus reached America, Copernicus enrolled in the Jagiellonian University in Cracow, one of the first universities in northern Europe to be influenced by the winds of Humanism blowing from Italy. By 1500 there were about eighty universities in Europe, a very different picture from the days of Roger Bacon. Cracow had a good tradition in astronomy: in particular, Albert of Brudzewo founded a school of astronomy and mathematics, and he may have influenced Copernicus's early instruction.

In 1496 Copernicus enrolled at the University of Bologna, Italy, to study canon law. By then, however, astronomy occupied much of his attention. He assisted the astronomer Domenico Maria de Novara, who was known for supporting the idea of the precession of the equinoxes. The notion that the Earth wobbled as it rotated may have influenced Copernicus's later decision to make the Earth move in an orbit. It is certain that he read extensively from many Greek classics, and that he was aware of Aristarchus's heliocentric model, which was quoted by Archimedes, Plutarch, and other writers readily available then. The printing press with movable type had been invented thirty years before Copernicus's birth, making books not only much cheaper but also much easier to find.

Europe had finally awakened from the long slumber of the Dark

Ages. Portuguese and Spanish sailors were stretching the limits of the known world, while Leonardo and Michelangelo were about to produce some of the greatest masterpieces of the Renaissance. Though, as the British philosopher Alfred North Whitehead, writing in 1925, somberly remarked, "By the year 1500 Europe knew less than Archimedes, who died in 212 B.C.," we can at least say that they were catching up fast.

In 1497 Copernicus made the first of his few astronomical observations, the passing of the star Aldebaran behind the Moon. In 1500, he gave a lecture in Rome about his observation of a partial eclipse of the Moon. By then Uncle Lucas had made him canon of the Frauenberg cathedral, a sort of administrator with a cushy salary and little work to do. In 1501 he returned to Italy, this time to Padua to study medicine, although he came back two years later with a degree in canon law from the smaller University of Ferrara, completing his strange academic career.

After spending a few years with Uncle Lucas as a kind of personal diplomatic secretary and physician, Copernicus finally settled at his job as canon of Frauenberg. He stayed there all his life, mostly isolated from society, glancing at the Baltic from the top of his gloomy tower. He was not fond of people, although he did live for a long time with a younger divorced woman, Anna Schillings, until the local bishop ordered him to let her go; it wasn't acceptable to have a canon living in sin in his own diocese. He had only one close friend, canon Tiedemann Giese, who would later play a crucial role in persuading Copernicus to publish his work in astronomy. If not for Giese's gentle but insistent prodding, this work could have been another of those hidden in the attic. As Arthur Koestler remarked, "Giese is one of the silent heroes of history, who smooth its path but leave no personal mark on it."

Being a canon left plenty of time for Copernicus to think about astronomy. Between 1510 and 1514, he composed a slim work summarizing his main ideas, the *Commentariolus*, "Short Commentary." He didn't publish the work, sending only a few manuscript copies to a select audience. Copernicus believed in the old Pythagorean ideal of brotherhood secrecy; only those initiated in the intricacies of mathematical astronomy were allowed to partake in his wisdom. This was certainly an odd position taken by someone educated in the heart of the Humanistic tradition. Was he gauging how "dangerous" his ideas were? Was he simply insecure about his results, wanting to avoid widespread criticism? The reasons for Copernicus's secrecy remain a topic of dissent among experts.

In the *Commentariolus* Copernicus states his basic conclusions:

1. The Sun is the center of the planetary system and thus of the Universe.
2. The Moon and only the Moon goes around the Earth.
3. The Earth spins about its axis.
4. The Earth and the planets orbit the Sun in circular orbits.

With this arrangement Copernicus put Aristotle's neatly divided cosmos into total disarray. If the Earth is not the center anymore, the division into the two realms of being (the Moon and above) and becoming (below the Moon) is demolished, along with the increasing decadence from the Empyrean sphere to Hell. The center of the Universe is not the devil, but the source of all light and energy, the keeper of life itself, "the visible god."

What led Copernicus to propose such a departure from traditional wisdom? One answer can be found in the beginning of the work, where Copernicus argues that Ptolemy's system of equants is not satisfactory, because it violates the Platonic requirement of *uniform* circular motion of all heavenly bodies.* He writes that Ptolemy's system "neither sufficiently achieved nor [is] sufficiently in accord with reason," and that "having become aware of these defects, I often considered whether there could perhaps be found a more reasonable arrangement of circles ... in which everything would move uniformly about its proper center, as the rule of absolute motion requires." By modifying Ptolemy's theory he was trying to "save the phenomena," faithful to Plato's doctrine of uniform motion. He was still looking backward, not forward.

But there is another reason why Copernicus was led to propose his new system, a purely aesthetic reason, based on the periods of revolution of the several planets. He placed Mercury closest to the

---

*Recall that with the equant, the center of the epicycle does not rotate uniformly around the center of the deferent, but around the equant point.

Sun, followed by Venus, the Earth, and then Mars, Jupiter, and Saturn, all surrounded by the sphere of the fixed stars. In justifying this arrangement of the heavenly spheres, he writes that

> [this ordering] is also in the same order [as] the speeds of the orb's revolution . . . so Saturn comes back to its starting point in thirty years, Jupiter in twelve years, Mars in two years, and the Earth in one yearly revolution. Venus carries out one revolution in nine months, and Mercury in three.

Copernicus's system naturally explained the differences in orbital period of the planets; the farther away from the Sun, the longer it takes for a planet to complete one revolution. There was, after all, a simple order to the cosmos. As he commented in his magnum opus years later, he found "a clear bond of harmony in the motion and magnitude of the orbs such as can be discovered in no other wise." A bond of harmony! Copernicus was a Pythagorean, searching for the "correct," or most harmonious, geometrical arrangement of the cosmos. No wonder he was so dissatisfied with Ptolemy's intricate system of equants and no obvious ordering of planetary orbits. Unfortunately, when he had to fit observational data into his model, all the simplicity was lost in a sea of epicycles and even "epicyclets," little epicycles attached to larger epicycles, a Copernican invention. Even the Sun was not quite at the center of the planetary orbits, but slightly displaced from the true center.

Was Copernicus disappointed by the complexity of his system? Not at all. In the last paragraph of the work he proudly announced that "altogether, therefore, thirty-four circles suffice to explain the entire structure of the Universe and the entire ballet of the planets." In his mind he had achieved his goal, which was to demonstrate how Plato's rule was compatible with a harmonious ordering of the cosmos. This is hardly the achievement of what we call a revolutionary. He had resuscitated the Pythagorean dream of two thousand years before. The Sun and planets were partners in a dancing Universe. The epicycles were merely the bricks of this grand geometric edifice.

In the *Commentariolus* Copernicus announced that all the details, i.e., geometric proofs that lent plausibility to his system, would be provided in a forthcoming work. Another thirty years would pass before the work was shown to the world, and only on the insistence of Giese and Copernicus's only pupil, Georg Joachim Rheticus (1514–1574). Why did it take him so long? Did he suffer religious persecution or peer criticism from the ideas advanced in the *Commentariolus*? Evidence points to the contrary: the *Commentariolus* didn't cause any great stir in the academic circles or harsh reprimands from his ecclesiastic superiors. If anything, Copernicus enjoyed a certain fame.

In 1514 he was invited to take part, with other astronomers, in a reform of the calendar. He refused, claiming that no reform would be satisfactory before the motions of the Sun and Moon were better known. In 1532 the personal secretary of Pope Leo X presented a seminar on Copernicus's work to a small audience in the Vatican gardens. His ideas must have been well received, because three years later Cardinal Schoenberg, who was quite close to the pope, urged Copernicus to "communicate your discoveries to the learned world" by publishing them. This is hardly the attitude of a Church interested in suppressing new ideas. In fact, the Church would be forced into an official position only later, due to Galileo's open challenge to a geocentric Universe based on Aristotelian dogma. But this occurred many years after Copernicus's death. Sure, people were being executed for witchcraft, Rheticus's father being one of the victims. However, proposing a new astronomical system was not viewed as a sign of having a pact with the devil, or of openly defying the Scriptures. One of the few criticisms directed at him came not from the Catholic Church but from Martin Luther, who remarked in an after-dinner joke that "there is talk of a new astrologer who wants to prove that the Earth moves and goes around instead of the sky. . . . The fool wants to turn the whole art of astronomy upside-down." Calling somebody a fool does not seem to indicate any real sense of threat. But Copernicus would not publish. At least not before the arrival of Rheticus.

In May 1539, a twenty-five-year-old professor of mathematics from the new Lutheran University of Wittenberg, full of Renaissance fire, came to visit a now aging Copernicus. Rheticus was in awe of the Copernican ideas, and wanted to make them public. It is interesting that this representative from the heart of the Lutheran domain would be allowed to visit a canon in a Catholic church; by this time the bishop had ordered all Lutherans to leave Ermeland.

Under the combined attack of Giese and Rheticus, Copernicus slowly gave in. In 1540, Rheticus published a summary of the work to be, called *Narratio Prima,* "the First Account." In it he forcefully put forward his opinion about the veracity of the hypotheses in the work, proclaiming Copernicus a prophet: "A boundless Kingdom in astronomy has God granted my learned Teacher. May he rule, guard, and increase it, to the restoration of astronomic truth." In May 1542, Rheticus delivered the lengthy manuscript of *De revolutionibus orbium coelestium,* "On the Revolutions of the Heavenly Orbs," to the hands of Petreius, a famous printer from Nuremberg. (This title reflects Copernicus's belief that the planets were attached to thick celestial spheres that carried them around the Sun.) But the drama was far from over.

Unfortunately, Rheticus could not remain in Nuremberg to supervise the printing of the manuscript. Rumors of homosexuality forced him to leave Wittenberg for another post in Leipzig, which he had to assume right away. He left the manuscript under the supervision of Andreas Osiander, an influential Lutheran theologian. Osiander and Copernicus had been exchanging letters for some time, in which they argued about the true meaning of astronomical systems of the world: Were they supposed to be real descriptions of the Heavens, or mere mathematical models, calculational tools, as in the days of Ptolemy? Osiander clearly sided with the latter view. In a letter to Copernicus dated April 20, 1541, Osiander wrote that these "hypotheses are not articles of faith but bases of computation, so that even if they are false, it does not matter, provided that they exactly represent phenomena." Copernicus didn't yield to Osiander's opinions. He dedicated his work to none other than Pope Paul III, expressing his view that Scripture should not be invoked to justify astronomy. Firm in his belief in the reality of the heliocentric hypothesis, he had finally overcome his fears. But the book was in the hands of Osiander. Without asking for Copernicus's consent, Osiander added an anonymous preface in which he maintained that all the models in the book were mere hypotheses, "which need not be true, or even provable." And that so far as hypotheses are concerned, let no one expect anything certain from astronomy, which cannot furnish it, lest he accept as the truth ideas conceived for another purpose [i.e., as calculational tools], and depart from his study a greater fool than when he entered it.

Paralyzed by a stroke in December 1542, Copernicus was unaware of this betrayal, or if aware, helpless to do anything about it. According to Giese, he saw the final version of the printed book, the embodiment of his life's work, only on May 24, 1543, the day he died. How desperate he must have felt when he realized the extent to which his thoughts had been corrupted, and how silent his lips had to remain.

Despite an action brought before the Nuremberg city council, Giese was not able to replace Osiander's preface. As a result, the next three generations of astronomers believed Copernicus wrote these words that so clearly contradicted his convictions. Not until 1609 would Johannes Kepler expose the hoax in his groundbreaking work on the orbit of Mars, his *Astronomia nova*, or "New Astronomy."

THE SEDUCTION OF SYMMETRY

In the peaceful little German town of Weil der Stadt, near the Black Forest, there lived, in a not so large house, the large family of mayor Sebaldus Kepler. He was a proud and powerful man, and even "eloquent, at least as far as an ignorant man can be," according to one of his many grandchildren, Johannes, born December 27, 1571. When he was twenty-six, Johannes wrote down astrological portraits of most members of his family, probably in a desperate attempt to purge himself from their pathological influence and to justify their sick temperaments through the malefic intervention of the stars.

Grandma Kepler was "restless, clever, and lying, of a fiery nature, an inveterate troublemaker, violent. . . . And all her children have something of this." Grandma and Grandpa Kepler had twelve children. The first three died in infancy. Next came Heinrich, Johannes's father, a cruel mercenary soldier who constantly beat his wife and children, a man "vicious, inflexible, quarrelsome, . . . a wanderer. . . . [In] 1577: he ran the risk of hanging [for some un-known crime]. [In] 1578: a hardjar of gunpowder burst and lacer-ated my father's face. . . . [In] 1589: treated my mother extremely ill, went finally into exile, and died." Of his other eight uncles and aunts, Kepler would write that Uncle Sebaldus was "an astrologer and a Jesuit . . . [who] led a most impure life. . . . Contracted the French sickness [syphilis]. Was vicious and disliked by his fellow townsmen . . . wandered in extreme poverty through France and Italy." Aunt Kunigund "was the mother of many children, poisoned they think, in the year 1581." Aunt Katherine "was intelligent and skillful, but married most unfortunately . . . now a beggar."

Of his mother, Katherine, Kepler writes that she was "small, thin, swarthy, gossiping, and quarrelsome, of a bad disposition." Raised by an aunt who was burned at the stake for witchcraft, Katherine barely escaped the stake herself. (If not for Kepler's intervention, she would have been burned alive.) She was known to cast evil spells on her enemies and to be an expert on herbal potions. Certainly, be-tween fact and fiction, young Johannes must have felt that he was cursed by the stars. Of his six siblings, three died in childhood, and two miraculously grew up to become fairly normal people. However, the one closest in age to Johannes, Heinrich, was an epileptic whose life was a parade of sufferings and misadventures.

What about young Johannes? His childhood was a succession of illnesses, beatings, and accidents. A weak, premature child, at four he almost died of a bout of smallpox that left his hands badly crip-pled. "When fourteen," he later wrote, "I suffered continually from skin ailments, often severe sores, often from scabs of chronic putrid wounds in my feet which healed badly and kept breaking out again. On the middle finger of my right hand I had a *worm*, on the left a huge sore. . . ." The list goes on, but I think you get the picture. It is the story of a sickly and depressed child, overwhelmed by a host of terrible circumstances beyond his control. Yet Kepler grew up to be-come one of the most productive and brilliant minds ever. Sur-rounded by pain and disease, he looked beyond in search of beauty and truth, purging himself through his creative power. In his work he searched for the inner peace that life so bitterly denied him.

Two memorable events marked Kepler's childhood. In 1577, when he was six, he was taken by his mother to see a shining new light in the skies, a long-tailed comet brighter than Venus. When he was nine, he recalls being "called outdoors by my parents especially, to look at the eclipse of the Moon. It appeared quite red." These events must have made an impression on Kepler, although his inclination toward astronomy would come only much later. Like many young men in those days, Kepler took advantage of the Protestant Church's need for a learned clergy to further his education. At thirteen, he joined a theological seminary, where he was exposed to the Greek classics, mathematics, and music. One would have thought that by leaving the accursed house in Weil der Stadt, Kepler would have felt a little better. Unfortunately, he carried much of that emotional heaviness within himself, and his relationship with the other boys at the seminary was terrible. He fought with everyone, was beaten many times, and was always getting into trouble. At seventeen he moved to the prestigious Lutheran University at Tübingen. Things only got worse, as Kepler's own teenage memories attest:

> February 1586: I suffered dreadfully and nearly died of my troubles. The cause was my dishonor and the hatred of my school fellows whom I was driven by fear to denounce. . . . 1587: On 4 April I was attacked by a fever from which I recovered in time, but was still suffering from the anger of my schoolmates with one of whom I had come to blows a month before. Koellin became my friend; I was beaten in a drunken quarrel by Rebstock; various quarrels with Koellin. . . . 1590: I was promoted to the rank of Bachelor. I had a most iniquitous witness, Mueller, and many enemies among my comrades.

But Kepler knew that the fault was often his. In one of his many piercing pieces of self-criticism, Kepler wrote of his "anger, intolerance of bores, an excessive love of annoying and of teasing." Writing in the third person, he would go so far as to compare himself to a dog:

> That man has in every way a dog-like nature. His appearance is that of a little lapdog . . . he liked gnawing bones and dry crusts of bread. . . . His habits were similar: He continually sought the good will of others, was dependent on others for everything, ministered

to their wishes, never got angry when they reproved him, and was anxious to get back into their favor.

At twenty Kepler graduated from Tübingen. He continued on his road to become a clergyman by joining the theological faculty. If not popular with his classmates, he was certainly popular with some of his teachers. Among them, Michael Mästlin became an important influence. Mästlin was one of the several astronomers who attacked the old Aristotelian doctrine of the division of the cosmos into two realms, after showing that the great comet of 1577 was certainly beyond the lunary sphere.* Possibly due to Mästlin's teachings, Kepler became a supporter of Copernicanism while still at Tübingen.

He was still pursuing his plans to become a clergyman, yet in 1594 he was recommended by his teachers in Tübingen to substitute for the teacher of astronomy and mathematics at the Lutheran school in Gratz, capital of the Austrian province of Styria. That meant not only the usual teaching duties, but also the title of "official mathematicus" of Styria. As such, he was supposed to compose an annual calendar of astrological forecasts. His first calendar was a success, predicting both a cold spell and a Turkish invasion. He soon became better known as an astrologer than as a teacher.

Kepler's attitude toward astrology was typical of a man in an age of transition. He constantly cast horoscopes for a living, sometimes despising this activity and sometimes confessing the irresistible attraction he felt toward it. He would alternatively write that astrology was "a dreadful superstition" and a "sortilegous monkey-play," but also that "nothing exists nor happens in the visible sky that is not sensed in some hidden manner by the faculties of Earth and Nature." Or "that the sky does something to man is obvious enough; but what it does specifically remains hidden." In other words, he believed in some hidden *physical* cause behind the supposed success of astrology. Innocent as this may sound, it represents a completely new approach to interpreting celestial phenomena. There is a hidden *cause* for the way things work, be it in astrology or in astronomy.

---

*Recall that according to Aristotle, change was only possible below the lunary sphere. Thus, comets were attributed to rare atmospherical phenomena. In fact, so were meteors, explaining why the study of the weather became known as meteorology.

In his search for this cause Kepler would bring physics into the study of the cosmos, creating a truly new astronomy.

The blinding insight that would change Kepler's life came unexpectedly, during one of his lectures. Before moving to Gratz, his interest in the Copernican system had been justified mainly by his mystical attraction to the idea of a Sun-centered Universe. Any other arrangement was completely nonsensical to him. Still, positing the Sun in the center didn't explain other cosmic mysteries, such as: Why were there only five planets plus the Earth orbiting the Sun? Why not ten or fifty? Was there some *secret symmetry* in the Universe that could explain this number? What about the relative distances between the planets? Was there a reason for them? Although Kepler's initial efforts to answer these questions were fruitless, they continued to bother him. On July 9, 1595, while lecturing on astrological conjunctions of Jupiter and Saturn to a handful of sleepy students, the solution suddenly came to him. Later, he wrote of this moment: "The delight that I took in my discovery, I shall never be able to describe in words."

The answer was geometry. Kepler knew that there were only five "Platonic solids," the only regular solids that can be built in three dimensions. They are "perfect" in the sense that all their faces are identical, as shown in the figure on page 80. Thus, a *tetrahedron* is made of four equilateral triangles; a *cube* is made of six squares; an *octahedron* is made of eight equilateral triangles; a *dodecahedron* is made of twelve pentagons; and an *icosahedron* is made of twenty equilateral triangles. No other closed solids in three dimensions can be made out of identical faces.

Kepler realized that he could explain not only the mutual distances between the planets but also their number by using the five Platonic solids. His clever idea consisted in nesting them, one inside the other like *matryoshkas*, the traditional Russian dolls. Being highly symmetric, it is possible to inscribe a sphere inside each of the solids, so that the surface of the sphere touches the inner faces of the solid. Similarly, it is possible to circumscribe a sphere around each solid, so that the inner surface of the sphere touches the outer surface of the solid. For example, see the nested triangles and squares in the left of the figure.

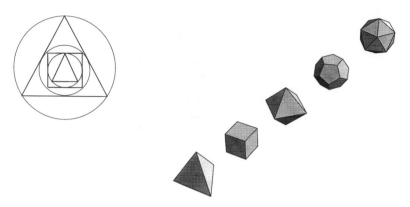

Nested triangles and squares and the five Platonic solids

By alternating sphere-solid-sphere-solid, etc., Kepler built a geometrical model of the Universe wherein the spheres represented the
orbits of the planets, while between two spheres there was a Platonic
solid. The distances between the planets were fixed by the way the
solids fit within each other. And since there were only five solids,
Kepler could fit only six spheres, explaining the number of the planets. In the center of the arrangement, the Sun reigned supreme.

Of course, we now know that in addition to the five planets visible to the naked eye there are three more, Uranus, Neptune, and
Pluto. But in Kepler's day the list stopped at Saturn. It is remarkable
that had Kepler known of the extra three planets, his scheme would
be useless. And yet his ignorance was his blessing; in trying to fulfill
his Pythagorean vision, he would later stumble onto the laws governing planetary motion, which, by the way, would work also for the
planets he didn't know existed.

The remarkable thing was that Kepler's scheme worked. Well, it
almost worked. It fit the Copernican model to about five per cent.
But the beauty and simplicity of the scheme, allied to its tremendous power to "explain" some very basic questions about the structure of the cosmos, captivated Kepler. Geometry was the key to
unlock the secrets of the Universe. The mystical Pythagorean tradition, once more, brought humans closer to the mind of the cosmic
Designer. And Kepler, the little lapdog, was the bridge. It was obvious, it was clear, it was the only possible solution. And, of course, it
was crazy.

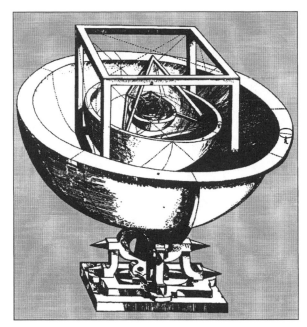

A physical model of Kepler's geometric universe

Although wrong, and physically unjustifiable, this vision of geometric order in the cosmos would dominate Kepler's thought for the rest of his life, the fundamental metaphor of his deeply felt rational mysticism. But his genius went beyond simply trying to revive old Pythagorean ideas. To Greek reasoning he added the most distinctive aspects of modern science. First, theories must conform to observations, and not vice versa. Second, theories describing phenomena must be physical, i.e., they must reveal the causes behind the observed behavior. This is where Kepler broke with the past. Carried by his cathartic vision of celestial harmony, he obtained the first mathematical laws describing planetary motions. His harmony was not purely aesthetic. It was quantitative, fed by observations.

After working frenetically for a few months, Kepler published in 1596 his first great work, the *Mysterium cosmographicum*, "The Cosmic Mystery." The first strong defense of the Copernican system, it appeared fifty-three years after the publication of "On the Revolutions." (The Copernican revolution was indeed quite slow to take off.) In this book Kepler stated that his five Platonic solids provided the skeleton of the cosmos, but he elaborated these ideas with a naive Christian mysticism and intuitive physical explanations for the causes of planetary motions. It was a book like none other ever written, a mixture of classic Greek philosophy, number mysticism, Christian theology, and rudimentary physics.

Kepler believed that the cosmos was an embodiment of the Holy Trinity. God was the Sun in the center, the Son was the sphere of the fixed stars, and the Holy Spirit was the power emanating from the Sun responsible for the motion of the celestial bodies, permeating the whole Universe. It was "a moving soul," the power of God to generate motion. He wrote that "either the souls which move the planets are less active the farther from the Sun, or there is only one moving soul in the Sun, which drives planets the more vigorously the closer the planet is." (He would later abandon the idea of soul for that of force.) The outer planets moved more slowly because this power "diminished in inverse proportion to distance as does the force of light." This is where physics enters the field of astronomy for the first time. Although Kepler's ideas were incorrect, the fact that he thought of searching for a cause is in itself revolutionary. With his brilliant intuition, he got very close to the concept of gravitational force.* He would get even closer in his next book, aptly titled *Astronomia nova*, "New Astronomy." But that was still thirteen years in the future.

On September 28, 1598, Catholic authorities representing the Counter-Reformation ordered all Lutheran teachers to leave Gratz. Kepler had some privileges by then, due to his growing fame, but was forced to start looking for another position. In February 1600, he arrived with his family in Prague, to work as an assistant to Tycho Brahe, the greatest astronomer of the time. It was a match made in hell. The only two things Kepler and Tycho had in common were arrogance and a passion for the stars. But they desperately needed each other. To understand why, we must backtrack a bit to tell the story of Tycho.

---

*He thought the force was confined to the plane of the planet's orbit, like spokes of a bicycle wheel connecting the center of the wheel to the rim. In fact, the force spreads in all directions.

THE ASTRONOMER PRINCE

The childless Joergen Brahe, vice-admiral to Frederick II, King of Denmark, so much wanted a son that he made his brother, the governor of Helsingborg Castle, promise that he could adopt his next male child. The governor agreed. In 1546, his wife gave birth to twin sons, but, tragically, one was stillborn. The governor refused to let the surviving boy, Tycho, go to his brother. Joergen was furious and ended up kidnapping his own nephew. After many threats and fights the governor finally gave up, knowing that at least Tycho would be raised with all the pomp befitting a true Brahe. That he certainly was.

Like most of his family members, Tycho was expected to become a diplomat or statesman. When he was thirteen, he was sent off to the Lutheran University of Copenhagen to study philosophy and rhetoric. Not long after he arrived, he had an experience that would change his life forever. In 1560 he observed a partial solar eclipse. What amazed Tycho was not just the beauty of the event, but the fact that astronomers had *predicted* when the eclipse would occur, that it was possible to know the course of the heavens with such certainty. From then on, his interest in astronomy would only grow, to the despair of his relatives.

When Tycho was sixteen, Uncle Joergen sent him to the University of Leipzig with a tutor, Anders Vedel, who was supposed to make sure that young Tycho would forget his foolish passion for the stars. But it was too late for that. After a year of constantly catching Tycho reading astronomy books or finding instruments hidden under his bed or in his closet, Vedel gave up on his mission. Tycho went on to several other universities until finally returning to Denmark in 1570. By then he had acquired a most distinctive facial trait: a gold and silver alloy that replaced part of his nose, which was sliced off in a duel.* With his huge balding head, small gleaming black eyes, and long twirled mustache, it is fair to say that Tycho didn't look like the kind of person you would want to have an argument with. His temperament more than matched his looks.

---

*It seems that the duel was the result of a dispute with one of his own relatives. The fight was to decide who of the two was the more talented mathematician, evidence that things haven't changed much in academic circles.

Unlike Copernicus or Kepler, who were driven into astronomy by more philosophical and mystical reasons, Tycho was a true observational astronomer. After the initial excitement of the solar eclipse, he quickly became aware that the data then available were far from accurate. His passion for precision knew no bounds. He was constantly developing new instruments that would enable him to measure with ever-increasing precision the position of heavenly bodies. In this he was fortunate that he could afford the huge costs associated with these large instruments, such as a quadrant of brass and oak thirty-eight feet in diameter.

Tycho put his money to good use by amassing a massive amount of data of unprecedented accuracy. Furthermore, he realized that positional measurements had to be not only accurate but also taken continually; if we are to understand the true path followed by the planets, we need to observe them as often as possible. For example, imagine I gave you a piece of paper with some points scattered on it, and told you that these were my observations of the planet Jupiter during the last three months. I would then ask you to infer from the picture what was the shape of Jupiter's orbit. Your job would be much easier if you had on the paper ninety points (daily observations) than if you had only twelve points (weekly observations).

Tycho's fame as an astronomer grew quickly. He was lucky to have lived in a time when the heavens were quite unsettled, shining with new and unexpected lights. On November 11, 1572, Tycho was coming back from his alchemical laboratory when he noticed a "new star" in the constellation Cassiopeia. A new star? But this was impossible, at least according to Aristotelian dictums. How could change ever occur above the lunar sphere? This was no ordinary star, either, for it was bright enough to be visible even during daytime. Using his instruments, Tycho measured its position with respect to neighboring stars until the "stella nova" faded from view the following March. His conclusions were quite clear. The new star was farther away than the Moon, and was not a comet, since it didn't have a tail and also didn't move. He wrote a book, *De nova stella,* detailing his observations and the design of his instruments.* For good

---

*Nowadays we would call this event a "supernova" event, a huge release of energy that happens when large stars approach the end of their regular life cycle.

measure, the book also contained astrological forecasts, poems, meteorological diaries, and related correspondence. Many other astronomers wrote about the "new star," but Tycho's measurements far surpassed all the others. The crack on the Aristotelian cosmos was getting deeper.

Five years later a new phenomenon appeared in the sky, the Great Comet of 1577, the same one that so impressed young Kepler. Tycho easily showed that the comet was at least six times farther away than the Moon, another heavy blow to the Aristotelians. Between the supernova and the comet, Tycho concluded that the Aristotelian spheres did not exist, driving planets and stars around their orbits.

This conclusion also ran counter to the Copernican model, since for Copernicus the celestial spheres were real. In addition, Tycho didn't like the heliocentric element of the Copernican model. Unable to detect any stellar parallax,* he declared that the immovable Earth had to be somehow in the center of the cosmos. Finally, Tycho was unhappy with the Copernican model on religious grounds, since it contradicted the Scriptures.

Tycho then proposed his own model, a hybrid between a pure geocentric Aristotelian model and the Copernican heliocentric model. He placed the Earth back at the center with the Sun in a circular orbit around it, as in Aristotle's model. But then he placed all the planets in orbit around the Sun, creating a lopsided model of the cosmos. Surrounding all of them was the sphere of the fixed stars, also centered on the Earth. An obvious consequence of the model was that the orbit of Mars intercepted that of the Sun around the Earth. The fact that this didn't bother Tycho shows the confidence he had in his observations. If there was anyone who could smash the heavenly spheres, Tycho Brahe was certainly the man for the job. Although praised by those desperate to save the Earth-centered cosmos at all costs, Tycho's model was destined to become the last whisper of the dying Aristotelian universe.

Tycho was by then considered one of the greatest, if not the greatest living astronomer. With his explanations of the supernova

*Recall our discussion of stellar parallax in Chapter 2 and how it proves the motion of the Earth.

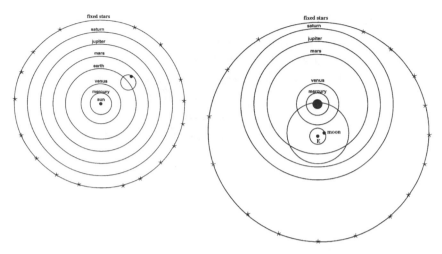

The cosmos according to Copernicus (left) and Tycho (right). To counter the heliocentric Copernican model, Tycho placed the Earth at the center. The rest of the planets revolved around the Sun, which in turn revolved around the Earth.

and the comet, he traveled around Europe in grand style, visiting many courts, showing off his instruments to potential patrons. King Frederick II, in a desperate attempt to keep Tycho in Denmark, made him an incredible offer. He gave Tycho the entire Island of Hveen, "with all the crown's tenants and servants who thereon live, with all rent and duty which comes from that . . . [for] as long as he lives and likes to continue and follow his *studia mathematices.*" Combining his extravagant character with his love for precision, Tycho built a large castle, the Uraniborg, or "Castle of the Heavens."

It was a remarkable place, not only for its powerful, fortress-like architecture, but also for what could be found inside. The huge construction was flanked by cylindrical towers with removable tops, where Tycho's instruments were stored. The library contained a large celestial globe made of brass, five feet in diameter, where Tycho and his assistants would engrave stars after they carefully measured their positions. There were countless galleries full of mechanical gadgets, including moving statues and hidden communications systems connecting different parts of the castle. The walls were adorned by drawings and epigrams from Tycho's own pen. In his main study he had the portraits of the eight greatest astronomers of all time, which included not only Tycho himself but also "Tychonides," his yet un-born descendant. In the basement Tycho had his own paper mill and printing press, an alchemical laboratory, and a dungeon, which he used to terrorize his tenants. He ruled over Hveen like a tyrant, intol-erant and arrogant with his servants, exuberant with his guests. In the manner of many kings, Tycho also had a "court" jester, a dwarf named Jepp, who could be found under Tycho's seat during ban-quets, talking incessantly while waiting for his master to toss him some scraps. Many illustrious visitors came to Uraniborg to partake in Ty-cho's feasts, including King James VI of Scotland. It was a research in-stitute like none else in history.

The bizarre idyll couldn't last forever. Uraniborg's exuberance started to decline after King Frederick's death in 1588. After a pe-riod of regency in which Tycho's brothers played an important part, Christian IV ascended to the throne of Denmark. Tycho's grandeur greatly displeased Christian, who was not willing to put up with his arrogance and tyranny. Tycho lost his favor with the king and in 1597 left Hveen with his entourage, complete with instruments, as-sistants, and the dwarf Jepp. After two years of wandering, Tycho ac-cepted an offer from Holy Roman Emperor Rudolph II to become his "Imperial Mathematicus." The offer, of course, included a lofty salary and Benatek Castle, near Prague. It was at Benatek that Kep-ler and Tycho finally met.

IN SEARCH OF COSMIC HARMONY

While Tycho was getting settled in his new castle, the expulsion of all Lutheran preachers and schoolmasters from Styria in 1598 fur-ther complicated Kepler's life. Even though he enjoyed some pro-tection from the widespread religious persecution, he knew he had to move soon. To complicate matters, Kepler by then was not alone. Bending to his friends' constant pressure, in April 1597 "under a calamitous sky," he married Barbara Muehleck, the daughter of a rich mill owner, twice widowed at twenty-three. Kepler would write that Frau Barbara was "simple of mind and fat of body," and had a "stupid, sulking, lonely, melancholy complexion," hardly the state-ments of a happily married man. He constantly complained of her ignorance, her lack of interest in his work, and her stinginess. In her defense, I imagine Kepler was not easy to live with, or to stir physical attraction. Apart from his scabs and the worms in his fingers, he is known to have taken only one bath in his whole life. And afterward he complained the hot water made him sick. The marriage lasted fourteen years, until Barbara's death at thirty-seven. Of their five children, only two survived. The first two died within two months of their birth, and Kepler's favorite son, Friedrich, died when still a little boy. Tragedy followed Kepler as obstinately as his own shadow.

Tycho was one of the few astronomers of the time who saw Kepler's genius beneath the foggy mix of speculations in the *Mysterium*. After years of excesses, Tycho's age was beginning to weigh on his shoulders, and he was badly in need of new assistants to carry on his work at Benatek. He had the most accurate collection of astronomical observations ever assembled, the building blocks of the new model of the cosmos, but lacked the talent of the architect to design the new edifice. In Kepler he secretly deposited his hope of seeing his efforts materialized. Previously Tycho had invited Kepler to visit him someday, in somewhat vague terms, but in December 1599, Tycho sent a formal invitation to Kepler, asking him to move to Benatek:

> I wish that you would come here, not forced by the adversity of fate, but rather on your own will and desire for common study. But whatever your reason, you will find in me your friend who will not deny you his advice and help in adversity, and will be ready with his help. But if you come soon we shall perhaps find ways and means so that you and your family shall be better looked after in future.

At Benatek the brilliant lapdog and the astronomer prince would spend eighteen months together, constantly arguing. Kepler continually demanded better treatment from the impervious nobleman, who was not as friendly as his letter implied. For his part, Tycho was initially reluctant to give his data to Kepler. How could he, the astronomer of kings, give the fruits of a lifetime of labor to the unknown plebeian at his side? Kepler would curse and bicker, as only he knew how. He would leave Benatek in an explosion of rage, only to come back, tail between legs, asking for forgiveness. Tycho knew he had no choice.

Finally, he found a problem to keep Kepler occupied: the orbit of Mars. This was no ordinary problem, as Tycho knew well. Mars's orbit, with its high eccentricity, is very treacherous. Kepler, with his usual lack of modesty, boasted he would take care of it in eight days. In fact, it took him almost eight years of hard work. But when he emerged from it, astronomy would never be the same again. The lapdog conquered the warrior planet and founded the new astronomy.

Tycho did not live to see his work immortalized. On October 13, 1601, he shared the table of the illustrious Baron Rosenberg. As was his custom, Tycho started to drink heavily early in the evening, but would not leave the table to relieve himself. Although his bladder was quite filled and distended, he would not stop drinking. For the next eleven days, he was taken by a high fever and had tremendous difficulty urinating. He died of acute uremia, poisoned by his own excesses. On his deathbed, Tycho kept repeating to Kepler, "Let me not seem to have lived in vain. Let me not seem to have lived in vain."

Two days after Tycho's death, Kepler was appointed the new Imperial Mathematicus, although his salary was a pittance compared to that of his predecessor. But Kepler was delighted. He immersed himself in his work, dividing his time between the challenge of Mars and the development of instrumental optics, another field he pioneered. Despite Kepler's efforts, however, Mars refused to cooperate. He initially tried to reinstate the equant in a Sun-centered model, merging the theories of Copernicus and Ptolemy. By this method he arrived at an agreement between model and observations to within eight minutes of a degree.* This was already far better accuracy than any model before him. But Kepler had enormous confidence in Tycho's data and knew he could do better. A true model of the skies had to fit the data.

Kepler was looking not only for the shape of Mars's orbit, but also for the physical causes of planetary motion. In the *Mysterium*, you will recall, he proposed a sort of "soul power" emanating from the Sun that drove the planets around their orbits like spokes of a

---

*A circle can be partitioned into 360 degrees. A degree, in turn, can be partitioned into sixty minutes. And a minute can be partitioned into sixty seconds.

bicycle wheel. Armed with new ideas coming from England, where the court physician to Elizabeth I, William Gilbert, had written a book exploring several aspects of magnetic phenomena, Kepler replaced the soul with magnetism as the primary force responsible for planetary motion. If, as Gilbert had shown, the Earth was a giant magnet itself, why not the Sun and all the other planets as well? In 1605 Kepler wrote,

> I am much occupied with the investigation of the physical causes. My aim in this is to show that the celestial machine is to be likened not to a divine organism but rather to a clockwork . . . , insofar as nearly all the manifold movements are carried out by means of a single, quite simple magnetic force, as in the case of a clockwork all motions [are caused] by a simple weight. Moreover I show how this physical conception is to be presented through calculation and geometry.

These are prophetic words. When Newtonian physics reached its climax during the eighteenth century, the Universe was indeed equated with a giant clockwork. It is this completely novel approach to astronomy that makes Kepler a truly revolutionary thinker. His achievements are even more remarkable when we realize that he had no predecessors. His methods may have been primitive, since he lacked the experimental approach being refined by Galileo at this same time. But his intuition was deadly. When he combined Tycho's precise data with his ideas about a central force emanating from the Sun, Kepler stumbled onto what we now call his "second law of planetary motion," namely that "The imaginary line connecting the Sun to the planet covers equal areas in equal times."

This law expresses the fact that, in a lopsided orbit, a planet moves faster the closer it is to the Sun; if the orbit were a perfect circle, the velocity would always be the same. The problem, though, was to find the correct description of the orbit. He toyed with all sorts of ideas, including epicycles, egg-shaped orbits, and other kinds of ovals. For Mars he produced fifty-one chapters, with hundreds of pages of calculations, and still couldn't crack the mystery. Kepler even seemed to take masochistic pleasure in providing his reader with detailed reports of all the wrong paths he took during his struggles with a problem. But if his writings don't make for light reading, they certainly provide invaluable insight into the elusive mechanisms of scientific creativity. Eventually, after initially discarding the idea as inconsistent with his magnetic hypothesis, he made the orbit an ellipse. And it worked! Thus was born Kepler's "first law of planetary motion," that "The planets travel in elliptical orbits with the Sun located at one focus of the ellipse."

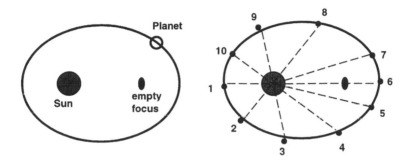

Kepler's first law of planetary motion states that planets travel in elliptical orbits around the sun, as on the left. According to the second law, if the numbers represent the planet's position at equal time intervals, the areas of the triangular segments are equal (right).

Kepler's two laws, immersed in a sea of calculations, appeared in his book *Astronomia nova* in 1609. The full title of the book reads, "A new astronomy *based on causation* or a *physics of the sky* derived from the investigations of the motions of the star Mars founded on observations of the noble Tycho Brahe." The book also contains Kepler's efforts to understand the tides and, of course, gravity. The volume was a magnificent effort to explain the motions of the heavens with one universal law, an achievement that was ultimately reserved for Newton. But the first steps were taken by Kepler. As the historian of science Gerald Holton writes, "His premonition of universal gravitation is by no means an isolated example of lucky intuition."

With the *Astronomia nova* Kepler had finally put the Sun in the true center of the Universe, the sole source of all planetary motions. But to Kepler, the Sun represented much more than that. It was the seat of God, His power emanating throughout the solar system. Kepler's system was not only heliocentric but also theocentric. As noted by Holton, the Sun had three perfectly complementary roles; as the *mathematical* center of planetary orbits, as the *physical* center insuring the continuation of the orbital motion, and as the *metaphysical* center, the temple of the Deity. Kepler's cosmos was a product of scientific creativity driven by a deeply religious instinct.

The publication of *Astronomia nova* was in itself a long battle, due to disputes with Tycho's heirs, which do not so much interest us here. After the work was finally in print, Kepler's life again fell into turmoil. The Emperor Rudolph's power and sanity started to crumble, and he finally abdicated on May 23, 1611, in favor of his brother Matthias. Kepler's six-year-old son, Friedrich, died the same year, and Barbara died early in 1612. The chaotic political situation and constant religious fighting made Prague a very dangerous place. In six years, it was to be the epicenter of the bloody Thirty Years War, the first all-European struggle for power, fanned by the religious rivalry of the Catholic and Protestant nobility. Kepler was once again stuck in the middle of the on-going religious tug-of-war which turned Europe into a vast battlefield. He left Prague for Linz, the capital of Upper Austria, where he spent the next fourteen years of his life. There he would finally concentrate once again on his lifelong obsession, the unveiling of the cosmic harmonies.

The *Harmonice mundi*, "Harmonies of the World," was completed in 1618. In it Kepler returns to the nested Platonic solids he introduced in the *Mysterium* twenty-one years earlier. The book is divided into five parts, the first two dealing with the concept of harmony in mathematics, and the following three with harmony in music, astrology, and astronomy. Kepler thus rescues the Pythagorean idea of harmony and dresses it in more powerful geometrical terms. Harmony manifests itself when our perception of order in nature is matched by simple geometrical archetypes, a resonance between the sensorial and rational experiences. To Kepler, it provides the unifying principle describing not only the motions in the heavens, but also the behavior of people, the changes in the weather, and the beauty of music. This all-embracing harmony is a manifestation of God's mind, as we perceive it in the world. It is, in short, the bridge between being and becoming.

Yet Kepler had to include his discovery of elliptical orbits plus all of Tycho's astronomical data. After much frustration Kepler arrived at a solution that deeply pleased him. He knew that in an elliptical orbit a planet's velocity increases when closer to the Sun. The key to the celestial harmony was in taking the ratio between the maximum and minimum orbital velocities. Kepler compared these numbers to the ratios of musical scales and obtained a very close agreement. Thus, Saturn corresponded to a major third, Jupiter to a minor third, Mars to a fifth, etc. He had finally unveiled the score of the celestial music heard by Pythagoras over two thousand years before! The composition became more complex and harmonic when he combined the speeds of different planets. The planets sang together, a motet praising the cosmic order. Kepler saw in the invention of polyphonic music an attempt to approach God:

> Man wanted to reproduce the continuity of cosmic time within a short hour, by an artful symphony for several voices, to obtain a sample test of the delight of the Divine Creator in His Works, and to partake of his joy by making music in the imitation of God.

Though Kepler insisted that the music was not for the ears, but for the intellect, the alluring power of celestial music inspired many seventeenth-century poets with its divine and unapproachable beauty. In Shakespeare's *Merchant of Venice,* we sense the frustration of having such music fall on deaf ears:

> How sweet the moonlight sleeps upon this bank!
> Here will we sit, and let the sounds of music
> Creep in our ears; soft stillness and the night
> Become the touches of sweet harmony.
> Sit, Jessica: look, how the floor of heaven
> Is thick inlaid with patines of bright gold:
> There's not the smallest orb which thou behold'st
> But in his motion like an angel sings
> Still quiring to the young-eyed cherubins;

> Such harmony is in immortal souls;
> But, whilst this muddy vesture of decay
> Doth grossly close it in, we cannot hear it.

In Milton's *On the Morning of Christ's Nativity,*

> Ring out ye crystal spheres,
> Once bless our human ears
> (If ye have power to touch our senses so)
> And let your silver chime
> Move in melodious time;
> And let the base of heav'n's deep organ blow,
> And with your ninefold harmony
> Make up full consort to th' angelic symphony.

To Kepler, though, finding the key to the celestial music did not complete the picture. The harmonic relations depended on the variations of the orbital speeds of the planets, but what about their distances from the Sun? If the Sun was the central seat of order in the solar system, then a relationship should exist between the planet's orbital period (the time it takes for a planet to complete one orbit around the Sun) and its distance from the Sun. Time and space should be somehow bound together by the power emanating from the Sun. After much trial and error, Kepler obtained his "third law of planetary motion" that "The square of the planet's orbital period is proportional to the cubic power of its mean distance from the Sun."* The law matched Tycho's data beautifully.

In this law Newton would later find the key to universal gravitation. It is to Newton's credit that he managed to distill Kepler's three laws from the confusing interweave of fancy and science typical of his writings. From a modern point of view, they are the essential part of Kepler's legacy. But to Kepler they were simply one of the many building blocks of his vast intellectual construction. His inspiration came not from the focused pursuit of particular laws, but

---

*Since the planet orbits the Sun in an elliptical orbit, its distance varies between a maximum separation, the aphelion, and a minimum separation, the perihelium. The mean distance is the average of the two.

from his obsessive belief that geometry was the common language of the human mind and that of God. This is a theme that still plays an important role in scientific creativity today, although "God" is usually understood as "nature."

With the completion of *Harmonice*, Kepler had arrived at the climax of his creative career. But he didn't stop here. In the last twelve years of his life, although always struggling against wars, disease, and personal tragedy, Kepler produced several works, two of them of major importance. In 1621, he finished his longest work, the *Epitome of Copernican Astronomy*, the most detailed exposition of astronomy since Ptolemy's *Almagest*. It was to become the most important text in astronomy for the next hundred years, even though it was immediately put in the Index of Forbidden Books by the Catholic Church. In 1627, after years of wandering from town to town avoiding battles and looking for a trustworthy printer, he managed to print the *Rudolphine Tables*, which included all Tycho's catalogue of 777 stars (increased to 1,005 by Kepler), as well as several tables and rules to predict the positions of the planets.

By then Linz was also being destroyed by the bloody revolts springing from the Counter-Reformation, and becoming uninhabitable for Protestants. Kepler had offers to go to Italy and to England but refused. "Am I to go overseas where Wotton [Lord Bacon's envoy] invites me? I a German? I who love the firm Continent and who shrink at the idea of an island in narrow boundaries of which I feel the dangers in advance?" After years of religious persecution, wars, and illnesses, Kepler's paranoia was more than justified.

He ended up going to Sagan, in Silesia (today mostly in southwestern Poland), under the patronage of Albrecht von Wallenstein, Duke of Friedland and Sagan, top general of the Holy Roman Emperor Ferdinand II. Kepler became a Lutheran mathematician in a Catholic court. His salary, badly in arrears for years, was to be paid by the duke, but of course it never was. Although his fame was great, his pocket was always empty. He spent two years there, eighteen months of which he again searched for a printer. His health, never his greatest asset, started to deteriorate. But Kepler wouldn't settle down, resembling, as Koestler has remarked, more and more the legendary Wandering Jew. (Curiously, in one of his hypochondriacal delusions, he wrote to Mästlin that he had a sore on his foot shaped like a cross, the mark of the Wandering Jew.) Wallenstein was dismissed by the emperor, and Kepler moved again, for the last time in his life. He went to Leipzig on a cheap old horse, and then to Regensburg to request his money from the insolvent imperial diet. After three days there he was taken by a high fever, which increasingly clouded his mind. According to a local witness, in his delirious state all he did was "point his index finger now at his head, now at the sky above him." He died on November 15, 1630, and was buried in a local cemetery. In a final twist of fate, his grave was later destroyed during the Thirty Years War, his remains scattered forever. But his epitaph, from his own pen, survived:

> I measured the skies, now the shadows I measure
> Skybound was the mind, earthbound the body rests.

4

THE PIOUS HERETIC

Man has weav'd out a net, and this net throwne
Upon the Heavens, and now they are his owne. . . .
— JOHN DONNE

O f the many conflicts between religion and science throughout history, none has received more attention than the clash
between Galileo and the Catholic Church during the first half of the
seventeenth century. The events that led to the famous trial of
Galileo by the Roman Inquisition in 1633 have inspired countless
debates among historians and theologians alike. Even as late as 1982
Pope John Paul II called for a deeper study of the issue "to remove
the barriers to a fruitful relation of science and faith that the Galileo
affair still raises in many minds." And finally, in 1992, John Paul II
officially reversed the Church's condemnation of Galileo.

The pope's open-mindedness was completely foreign to the
Church of three centuries before. In those days, as we have seen in
the last chapter, the schism between the Catholic and Protestant
faiths was engulfing Europe in bloody conflict. The medieval supremacy of the Catholic Church was crumbling fast, together with
its Aristotelian view of the world. These were critical times for the
survival of the Catholic Church, and no challenge to its power was
to pass unanswered. The Inquisition was operating at full force,
spreading fear in the minds of those who dared contradict the word
of the Holy Fathers. As an additional weapon against the mounting

heresy, the Church in 1540 organized its own militia, the Society of Jesus, or Jesuit order, which was to be engaged in the spreading of Catholicism across the world. As for matters of Christian theology, the Council of Trent (1545–63) stipulated that no interpretation of the Scriptures differing from that sanctioned by the Holy Fathers was to be tolerated.

The burning of philosopher Giordano Bruno in 1600 is a sad example of the Church's war against religious deviation. Bruno's problems with the Church were caused more by his theological heresies than by his belief in an infinite Universe, populated by infinite suns like ours. His cosmological speculations and support for Copernicanism were indeed far ahead of his time and mark a true departure from medieval thought. But to the Church, his views on transubstantiation, the Trinity, and the substantiality of the human soul were far more dangerous. Bruno died an impenitent apostate, forever a symbol of the courage of the human spirit against blind censorship.

Although Galileo is commonly represented as one of the greatest martyrs in the fight for freedom of expression, and the Church as the intolerant villain, the truth is more complex. When Galileo launched his personal crusade to disprove the geocentric model of the cosmos, the Church's position concerning the arrangement of the heavens was somewhat flexible. Copernicus's book was not put in the Index until 1616, over sixty years after it was published, and even then it was not forbidden, but merely "corrected." A few sentences stating the heliocentric system as certain (as opposed to a mere hypothesis) were removed, as well as references to the Earth's being a "star."

Only when Galileo openly challenged the hegemony of the Church did the conflicts start that eventually led to his trial by the Inquisition. Convinced by the strength of his remarkable astronomical findings, Galileo declared the geocentric Ptolemaic model of the Universe as untenable. Moved by a combination of blind ambition and sincere piety, Galileo saw himself as the guiding light of the new Church. He would not only tell the professors of philosophy how stupid they were (he used the word *stupid* many times when referring to Aristotelians), but also tell the Holy Fathers how to interpret the Scriptures. This affront the Catholic Church could not tolerate.

The conflict between Galileo and the Church is an excellent, if tragic, metaphor for the clash between the old and the new. The blind arrogance of youth is stymied by the lack of flexibility of the old; the impatient ambition of the young not only stirs the old's fear of new ideas but poses a threat to its established power. In the short term the old guard wins, but if the strength of the progressive arguments is truly great, it invariably triumphs in the end. Although the Church did manage to silence Galileo, that victory was short-lived. A few decades after Galileo's death, Isaac Newton developed a new physics unifying the motion of heavenly bodies with the motions of objects on Earth, which carried the Sun-centered Copernican hypothesis into its logical conclusion. The transition from the old to the new was completed. The medieval Universe finally collapsed under the overwhelming strength of the new science, a testimony to Galileo's courage and genius.

THE MESSAGE FROM THE STARS

Galileo Galilei was born in 1564, the same year as Shakespeare, and seven years before Kepler. It is unfortunate that the two men who contributed so crucially to the emergence of a new worldview had such scant contact. They did exchange a few letters, but never really engaged in a debate over their ideas. While Kepler would write to Galileo praising his theories and discoveries, Galileo would write back mainly to complain of his adversaries and their plotting against him. It seems that although Galileo recognized Kepler's genius, he was put off by his style, which often buried the crucial issues under foggy arguments. And it is fair to assume that Galileo was not too keen on praising someone other than himself. The heavens were his exclusive property, and he was not prepared to share them with the German lapdog. To the end of his life Galileo would not endorse Kepler's elliptical orbits, preferring instead to adopt a simplified version of the heliocentric Copernican model.

Influenced by his father, Galileo in 1581 joined the medical school at the University of Pisa. He never completed his studies, and in 1585 went back home to pursue his growing passion for the study of "natural philosophy." During this time he was mainly concerned with the study of motion, to which he would make crucial contributions.

In 1582, or so legend tells, he was listening to mass in Pisa when he noticed the swinging motion of a large candelabrum, after an attendant pulled it aside to light the candles. Using his pulse to measure time, he noticed that although the swinging became less pronounced as the candelabrum slowed to a stop, the time between consecutive swings (the "period of oscillation") did not change by much. He then went home and repeated this experiment, using a string and a stone in place of the large candelabrum.* To his surprise, he observed that although the period of oscillation depended on the length of the string, being longer for a longer string, the period was *independent* of the weight of the stone! A heavy stone oscillates with the same period of a much lighter one. This contradicted the Aristotelian idea that heavier objects come faster to their state of rest, following their natural tendency to fall toward the Earth. For we can easily think of pendular motion as a periodic fall toward the final position of vertical rest, the closest position to the Earth. Galileo may have then, if not earlier, realized that something was seriously wrong with Aristotelian physics.

To this story we can add the famous one concerning the dropping of objects from the Tower of Pisa. Galileo's pupil and first biographer, Viviani, claimed that the incident indeed took place, in front of an incredulous crowd of Aristotelian professors and students. But since there is no official record of the experiment, historians are still divided as to whether it really happened. Legend or not, the story tells of Galileo dropping objects of different weights from the top of the tower to show that they would reach the ground at essentially the same time. The differences in the arrival times were mainly due to the friction between the air and the falling objects, which, of course, is sensitive to their shapes; a flat piece of paper falls much slower than if it is crumpled into a ball. But in a cylinder evacuated of all air, it will take exactly the same time for a feather and a coin to touch the bottom, if dropped from the same

---

*Here you have an example of a physicist's mind at work. The string and stone combination can perfectly mimic the swinging motion of the large candelabrum with the glowing candles. Galileo didn't have to go back to the cathedral and spend time convincing the priest that he wanted to study the details of "pendular motion" by swinging the large candelabrum. A good experiment distills a complicated phenomenon to its bare essentials, a crucial step when we attempt to describe its results in terms of a mathematical model.

height. This is a classic lecture room demonstration that never fails to cause some ooohs of amazement.

Galileo's reputation was becoming widespread, and in 1589 he joined Pisa as a lecturer in mathematics. In 1592 he was given the chair at the University of Padua, a prestigious position for a young man of twenty-eight. Galileo's style was completely novel for his time. Against blind obedience to Aristotle's ideas, he offered experimentation: Don't trust authority, trust reason, check it yourself before taking sides on a scientific issue. Experimentation for Galileo consisted of both real experiments and "thought experiments," those ones of hard (or even impossible) execution that take place inside the physicist's mind, which have proved to be a crucial tool in the development of theories. Later, he was to go beyond mere experimentation to obtain the *mathematical* relations that described the motion of falling objects. Galileo firmly believed that mathematics is the language of nature, and the world is constructed in such a way that the mathematical relations describing a phenomenon are the simplest possible.

In his emphasis on experimentation, combined with a mathematical description of the results, Galileo was the first truly modern scientist. His pioneering approach to the study of motion was crucial to Newton's formulation of the laws of mechanics and gravitation.

After these initial results, however, Galileo was to become silent on the physics of motion until nearly the end of his life. The reason for this prolonged silence was the arrival of an incredible invention, the telescope. Although there is still some debate about precisely who invented the telescope, the first license to build telescopes was obtained by the Dutch optician Johannes Lippershey on October 2, 1608, although there were reports of optical magnifying tubes at a Frankfurt fair in September. The instruments attracted so much attention that by April 1609 one could buy them in Paris. Galileo heard of the invention and quickly manufactured his own telescope, of better quality than the ones available. Being an astute social climber, on August 8, 1609, he invited the Venetian senate to examine the instrument from the top of San Marco's tower, stressing the importance of the telescope as a defensive weapon against an invasion by sea. It was a great success. The senate was so grateful that it

not only made his appointment in Padua permanent but also doubled his salary. (Even Galileo had to wait for tenure, although most
people don't get this kind of raise with promotion.) Apart from improving his financial and professional status, the telescope was to
become Galileo's greatest tool in his personal crusade against the
Aristotelian worldview; the skies would never be the same after
Galileo aimed his telescope at the stars.

Galileo discovered that there were many more stars than those
visible to the naked eye. Pointing his telescope at the constellation
of Orion, he counted at least eighty stars around the famous three
that we associate with Orion's belt. The Milky Way, he wrote, was
just an extremely dense agglomeration of stars. The Moon was far
from being a perfect sphere, as it was punctuated by mountains and
valleys, looking very much like the Earth. He went even as far as to
compute the height of the mountains, which he showed could measure as much as twelve thousand feet. He also conclusively showed
that the Earth provided a "secondary" source of light to the shad-
owed regions of the Moon, just as the Moon can brighten our dark
nights with its ghostly light. In short, Galileo proved that the Moon
was essentially like the Earth, to the horror of the Aristotelians.

That was only the beginning. Pointing the telescope at Jupiter,
Galileo noticed that the giant planet was not alone. Dancing around
it he saw four new "planets, never seen from the very beginning of
the world to our own times." These were four of the moons of
Jupiter, which Galileo hurried to call the "Medicean Stars," in a
clear gesture to impress Cosimo II di Medici, the Grand Duke of
Tuscany. This discovery was of tremendous importance. Rather than
being absurd, the hypothesis that the Moon revolves around the
Earth as the Earth follows its path around the Sun was proven correct. If Jupiter had its own satellites (a term coined by Kepler, who
marveled at Galileo's discoveries), why not the Earth? And if so, in
what sense was the Earth different from the other planets? Again,
the special status attributed to the Earth by the Aristotelians was seriously threatened.

In March 1610, Galileo published his discoveries in a booklet
called *Siderius nuncius*, translated variously as "The Starry Messenger," "Message from the Stars," or "Messenger from the Stars." This
polemic is relevant because the interpretation of the book's title can

be quite important in understanding Galileo's role as the self-assigned savior of the Church. Galileo believed he had exclusive access to the truths written in the skies. In his book *Il Saggiatore*, "The Assayer," published in 1623, Galileo wrote that "it was granted to me alone to discover all the new phenomena in the sky and nothing to anybody else. This is the truth which neither malice nor envy can suppress."

The book was a ruthless attack on the Jesuit priest Horatio Grassi, who dared confront Galileo on a series of disputes concerning comets and the heating of flying objects by friction, but the important point for us now is Galileo's belief that *he alone* had been "granted" the access to the phenomena in the sky. Although his statement was certainly blurred by his blinding ambition and hatred for his competitors, the question still remains as to *who* had granted him this privilege. Is it possible that Galileo believed himself to be the "Messenger from the Stars"? That he believed he had been chosen by God to provide the Church with the new truth concerning the skies? Although this point is speculative, seen from this angle the conflict between Galileo and the Church gains a new dimension.

Messenger from the Stars was a great success, bringing Galileo widespread fame and recognition. In September, after receiving an enthusiastic endorsement from Kepler, Cosimo named Galileo his "Chief Mathematician and Philosopher," adding a professorship at the University of Pisa as an extra bonus. Thanks to his clever manipulation of aristocratic patronage, Galileo ascended to the heart of the Tuscan court. In the following spring he went to Rome and was showered with the highest honors of the day. He was made a member of the prestigious Accademia dei Lincei, which feted him with a banquet. Pope Paul V received him in a friendly audience, despite his personal dislike for scientific matters. The Jesuit Collegio Romano, the center of intellectual activity of the Church, honored Galileo with celebrations that lasted an entire day. Galileo had the support of the chief astronomer of the Collegio, Father Clavius, who confirmed Galileo's discoveries to the head of the college, the Lord Cardinal Bellarmine, the greatest theologian of the day. Things couldn't have been going any better. Or could they?

THE LETTER TO CHRISTINA

Galileo's astronomical discoveries and his high position at the Tuscan court brought him tremendous prestige and widespread reputation. But not everyone was celebrating his accomplishments. A colleague once told me that in academic circles, as soon as you jump ahead of the pack, someone will be ready to knife you in the back. I would add that this serves only to prove that academic circles are just like any other professional circles, that scientists and scholars share all the good and bad qualities of being human. But it is common sense that if you do indeed become successful, you should at least try to keep a low profile. Galileo would have none of that. Back in Tuscany from his Roman triumph, he quickly got involved in a series of disputes with several professors of philosophy and, ominously, with the Jesuits.

In the spring of 1613, Galileo published his *Letters on Sunspots*, where he correctly argued that they were either on, or very close to, the Sun, contradicting the opinion of Father Scheiner, a Jesuit who claimed the spots were made of several small planets circling the Sun. At issue was the Aristotelian idea that the Sun was not subject to change, and hence could not display these scars that moved and changed shape quickly enough to be perceived by human eyes. Galileo also sought to make sure everyone knew that he had been the first to observe the sunspots, even though that was not at all clear. More important, Galileo presented in the *Letters on Sunspots* his first published support for Copernican ideas. By then he had added new observations to the ones published in his *Siderius nuncius*, which offered compelling (but not conclusive) evidence for the heliocentric model. Chief among these were his observations that Venus, like the Moon, has phases. This is possible only if Venus orbits the Sun, ruling out the Ptolemaic system.* The choice was now between the Copernican and the Tychonic system, which, as you remember, had all planets orbiting the Sun but the Sun still orbiting the Earth. Most Jesuits leaned toward the Tychonic system, since

---

*Recall that in the medieval arrangement, Venus's orbit is closer to the Earth than the Sun. Thus, it would never be possible to observe a full phase, since this requires that the Sun be between the Earth and Venus.

this avoided a potentially embarrassing reexamination of the Scriptures, but as a whole, the Church still didn't have an official position. Even after Galileo's support for the Copernican system was made public, he received several letters from high Church officials expressing their admiration for his work, including one from Cardinal Maffeo Barberini, soon to become Pope Urban VIII.

Galileo's problems started when one of his disciples, Father Benedetto Castelli, was invited for a dinner at the Tuscan court. Castelli had recently been appointed professor of mathematics at Pisa through Galileo's influence. As soon as he took his post, he received orders from the head, Arturo d'Elci, one of the Aristotelians with whom Galileo had crossed swords before, not to teach the notion of an Earth in orbit to his students. At the head of the table sat the Grand Duchess Christina of Lorraine, Cosimo di Medici's mother. She had also invited Dr. Boscaglia, a professor of philosophy.

The Grand Duchess was quite interested in the latest astronomical discoveries, in particular those new "stars" that carried the family name in faithful orbits around Jupiter. "Are they real or a mere illusion?" the Duchess asked. Both Boscaglia and Castelli promptly assured her they were real. After whispering something into the duchess's ear, Boscaglia went on to say that all of Galileo's discoveries were true, but that he was disturbed by the idea of the Earth in motion, since this contradicted the Scriptures. It is not clear what happened after this, but as Castelli was leaving the palace, he was summoned back. The duchess then resumed the argument that Galileo's discoveries contradicted Scripture, forcing Castelli to play the role of theologian, as he struggled to show that there were no contradictions. In Castelli's own words, "Only Madame Christina remained against me, but from her manner I judged that she did this only to hear my answers. Professor Boscaglia never said a word."

Castelli reported the incident to Galileo in a letter dated December 14, 1613. By the twenty-first, Galileo had written an elaborate answer, known as "Letter to Castelli," in which he cleverly tried to argue that the motion of the Earth does not contradict the Scriptures, *if only they were interpreted correctly.* A year later, he prepared a revised version that became known as the "Letter to the Grand Duchess Christina." His intentions were clear. He wanted the letter to become public property in order to silence once and for all the mouths of those who still used the Bible as an astronomy text:

> Let us grant then that theology is conversant with the loftiest divine contemplation, and occupies the regal throne among the sciences by this dignity. But acquiring the highest authority in this way, if she does not descend to the lower and humbler speculations of the subordinate sciences and has no regard for them because they are not concerned with blessedness, then her professors should not arrogate themselves the authority to decide controversies in professions which they have neither studied nor practiced.

He then goes on to say that propositions which have been proven beyond doubt must be distinguished from those which are still speculative. If propositions that have been proven (scientifically, that is) to be correct contradict the Scriptures, then their interpretation must be revised; the Bible does not err, but its human interpreters may.

Galileo underestimated the strength of his opponents; not only their power, but, as usual, their intellectual abilities. Deep down he knew he had no convincing proof of the Copernican hypothesis, just accumulated evidence. But to his mind the evidence was so compelling that now it was the Church's job to prove *him* wrong.

Meanwhile, opposition to Galileo's views was growing. In December 1614, a young Dominican monk, Tommaso Caccini, who had a reputation of being a troublemaker, preached a sermon in the church of Santa Maria Novella in Florence. The sermon was a deliberate attack on the Copernican system and caught the attention of another Dominican, Father Niccolo Lorini, who had earlier expressed dislike for Galileo and his ideas. Lorini then brought a copy of the "Letter to Castelli" to his fellow brethren at St. Mark's in Florence for discussion. The contents must have stirred up a lot of anger, for Lorini quickly proceeded to send a copy of the letter to the Inquisition in Rome early in 1615.

Galileo had friends of his own in Rome. During 1615 he exchanged letters with Cardinals Dini and Ciampoli, who kept him informed of the secret proceedings of the Inquisitors. He asked Dini to show his own copy of the "Letter to Castelli" to the Inquisitors, because he was worried it had been corrupted by Lorini on the way

to Rome, which indeed it had. In February, Ciampoli wrote to
Galileo that Cardinal Barberini

> would like greater caution in not going beyond the arguments used
> by Ptolemy and Copernicus [i.e., as mathematical hypotheses only]
> and finally in not exceeding the limitations of physics and mathe-
> matics. For to explain the Scriptures is claimed by theologians as
> their field, and if new things are brought in, even by an admirable
> mind, not everyone has the dispassionate faculty of taking them
> just as they are said.

Even Barberini, a supporter and admirer of Galileo, was telling him to
lie low.

Behind the scenes lurked the giant shadow of Cardinal Bellarmine, Master of Controversial Questions of the Roman College.
His reputation was such that, while he lay dying in his bed, an un-
ending procession of cardinals and other churchmen invaded his
chamber to touch and kiss his body. At his funeral, the authorities
had to struggle to keep the mob from tearing his body apart, so eager they were to collect relics, a scene that invokes visions of medieval hysterics.

Bellarmine spent his life fighting the Protestant heresy. He
would not be told by a mere mathematician, even a brilliant one
such as Galileo, how to interpret the Scriptures. His unofficial position was made clear when he received a letter from a Carmelite
theologian, Paolo Foscarini, on Copernicanism and the Scriptures.
Foscarini believed in the Copernican system and tried to show,
much as Galileo had in his "Letter to Castelli," that the two were not
incompatible. Bellarmine cleverly seized the opportunity to reply
not only to Foscarini, but also to Galileo. He summarized his
position in a short reply to Foscarini, making sure that Galileo saw
a copy.

Here Bellarmine displayed his tremendous political skills. He
started by commending Foscarini and Galileo for doing something
they had not done, i.e., insuring that the Copernican ideas are merely
hypotheses. To affirm that the Sun is the center of the Universe and
the Earth rotates around it, he wrote, "is a very dangerous attitude
and one calculated not only to arouse all Scholastic philosophers and

theologians but also to injure our holy faith by contradicting the Scriptures." Second, he added, "as you know, the Council of Trent forbids the interpretation of the Scriptures in a way contrary to the common agreement of the Holy Fathers." Third,

> if there were a real proof that the Sun is in the center, that the Earth is in the third sphere, and that the Earth goes around the Sun, then we should have to proceed with great circumspection in explaining passages of Scripture which appear to teach the contrary, and we should rather have to say that we did not understand them than declare an opinion to be false which is proved to be true. But I do not think there is any such proof since none has been shown to me.

In one move Bellarmine protected Galileo from the possible consequences of his pretentious acts (commending him for doing what he hadn't done), advised him to avoid interpreting the Scriptures (as determined by the Council of Trent), and left the door open for the case in which a *real* proof for the Copernican system did come up in the future (he realized Galileo had no such proof).

Galileo wouldn't listen. He decided to elaborate his case as stated in the "Letter to Castelli" with the famous "Letter to Christina," which he hoped Bellarmine would read. Against the advice of the Tuscan ambassador, in December 1615 Galileo went to Rome to try to clear his name.

We can only speculate why he chose to make such a dangerous move. Perhaps, being overconfident of his intellectual abilities, he thought his arguments would prove irrefutable. He was used to humiliating his opponents in oral disputations. Perhaps he was moved by a pious desire to save the Church from future embarrassment. Or perhaps arrogance and piety combined in his mind; he thought it was his mission to be the messenger of the new science to a Church firmly anchored in the past. Whatever his reasons, Galileo knew that by coming to Rome he would be creating the conditions for an open conflict, in which the authority of the Church was at stake.

Galileo didn't go to Rome empty-handed. He believed he had his "proof" of the motion of the Earth, which Bellarmine required as a necessary step toward the acceptance of the Copernican hypothesis. The proof consisted of his theory of the tides. The idea was simple, and quite wrong. The Earth both orbits around the Sun and rotates around its axis. According to Galileo, these two motions combine to make the ocean slosh around at a different speed from the land, causing the tides. This is roughly how he reasoned: Consider a city located by the ocean such as Rio de Janeiro, Brazil. As indicated in the figure below, at midnight both orbital motions point in the same direction, causing the land to "rush" forward, leaving the ocean behind; this is, according to Galileo, the low tide. At noon, the two motions are in opposite directions and the land moves slowly, while the water piles up, causing a high tide. To the obvious criticism that oftentimes there are more than one high and low tide per day, and that they occur at different times, Galileo would answer that these were details that can be attributed to secondary causes, such as the difference in depth of the ocean floors, the shape of the land bordering the waters, etc.

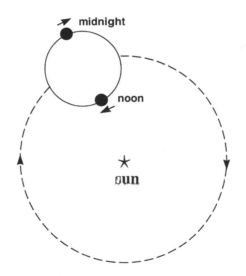

Galileo's theory of the tides. At midnight the orbital motion of the Earth coincides with its movement around the sun, causing the land to "rush" forward and leaving the tide behind. At noon, it reverses direction and the tides catch up.

The main problem with Galileo's argument is that he separates the motion of the land from that of the water, as if they obeyed different physical laws. To Galileo, all heavenly bodies obeyed a kind of "circular inertia," which explained why they could stay in circular orbit forever, without any force being applied to them.* On the other hand, the oceans were subject to the laws of motion of objects on Earth. Contrary to Aristotle, who claimed that the natural state of an object is to be at rest, Galileo realized that motion with a constant speed is as natural as rest. This is what we call *inertial motion.* To convince yourself that this is true, imagine being in a car that is moving with constant velocity on a perfectly smooth and straight road. All windows are covered, so that you can't see outside. In this case, you can't tell if the car is moving or if it is at rest; the two states of motion are perfectly equivalent.

Galileo tried to convince his friends to convey his ideas to the pope. Since no one was willing to do this, he finally convinced a young cardinal, Alessandro Orsini, to do the job. The pope was not pleased. According to the Tuscan ambassador, as soon as Orsini left the Holy Office, the pope called for Bellarmine. They quickly decided that Galileo's opinion was erroneous and against the teaching of the Holy Fathers. On February 19, 1616, the Qualifiers of the Holy Office were requested to give their opinion on two propositions: (1) that the Sun is the center of the world and wholly immovable; and (2) that the Earth is neither the center of the world nor immovable, but it both moves itself and also displays a daily, or diurnal, motion. The first proposition was declared "foolish and absurd, philosophically and formally heretical inasmuch as it expressly contradicts the doctrine of the Holy Scripture in many passages." The second proposition was declared to "deserve the like censure in philosophy, and as regards to theological truth, to be at least erroneous in faith." This was strong wording, and could have led to dangerous consequences to Galileo. However, this verdict was not immediately released; by March 5, 1616, a milder version was issued by the Congregation of the Index. One senses the influence of Bellarmine behind this decision.

While the Congregation of the Index condemned and prohibited Foscarini's book, and any other work that taught the Coper-

---

*Recall that at that time there was no concept of gravitational force, apart from Kepler's frustrated attempts, which Galileo neglected anyway.

nican system as true, Copernicus's *De revolutionibus* was not prohibited but only suspended, until corrections were made that erased any mention of his ideas as more than a hypothesis. There was no mention of the name Galileo anywhere in the decree. He had been admonished orally by Bellarmine, following the pope's request. The interesting point is that, according to the Inquisition files, this happened on February 25, two days after the Qualifiers finished their examination of the two propositions, but *before* the Congregation issued its decree; it seems that, as conjectured by the historian Richard Westfall, a compromise was forged between Galileo and Bellarmine. The Inquisition files state that Galileo was admonished by Bellarmine to

> abandon the said opinion; and in case of his refusal to obey, that the Commissary is to enjoin on him, before a notary and witnesses, a command to abstain altogether from teaching or defending this opinion and doctrine and even from discussing it; and, if he does not acquiesce therein, that he is to be imprisoned.

Galileo complied. However, rumors were circulating claiming that he was humiliated and punished by the Inquisition. By Galileo's request, Bellarmine issued a document stating that he had not been forced to abjure, nor been punished in any way; only that he had been notified of the contents of the declaration made by the Congregation of the Index. Galileo could go home with his head upright, but with his mouth shut. The first battle was over.

THE FINAL CONFRONTATION

For seven years Galileo wrote nothing. His next book was *Il Saggiatore*, in which he engaged in bitter polemics with the Jesuit Horatio Grassi, as we discussed above. By then Bellarmine was dead, and the balance of power was rapidly shifting. In 1623 Cardinal Maffeo Barberini, who seven years earlier had played an important role in smoothing things out for Galileo, became Pope Urban VIII. This was the opportunity Galileo was waiting for to launch a renewed attack on the Earth-centered universe of the Church. He dedicated *Il Saggiatore* to Urban and was received by him for six long audiences during the spring of 1624. The pope's admiration for Galileo was sincere. In 1620 he had written a poem to Galileo titled *"Adulatio Perniciosa,"* which has been translated as "Perilous Adulation," a very intriguing title. During Galileo's visit he gave him a silver and gold medal, a pension for his son, and a glowing letter to the Tuscan court in which he wrote of all the virtues "of this great man, whose fame shines in the heavens, and goes on earth far and wide."

Galileo asked Urban for permission to write a new book, a dialogue in which he would pit the Copernican and Ptolemaic systems of the world against each other. Galileo may have convinced Urban that his theory of the tides was indeed definitive proof for the motion of the Earth. The pope agreed, but insisted that the text make clear that God, through a miracle, could make the tides flow daily even on a stationary Earth. In other words, even though the Copernican hypothesis may best explain the phenomena, it is not possible to rule out that the hand of God is the ultimate cause of what we observe.

Galileo finished the original version of his *Dialogue on the Flux and Reflux of the Tides* in January 1630. The *Dialogue* was written as a free discussion among three men, Salviati, Sagredo, and Simplicio. Salviati was Galileo's spokesman, while Sagredo was supposed to be an informed and presumably neutral participant, who usually sided with Salviati. Simplicio was the Aristotelian, always being cleverly outwitted by Salviati. Although Galileo claimed that the name Simplicio was inspired by Simplicius, the sixth-century commentator on Aristotle, the sarcasm of the choice was obvious. The action takes place over four days, during which the Copernican and Ptolemaic systems are made to clash from all different angles. The first day is dedicated to the Aristotelian view of the cosmos, such as the division of the world into the two realms of being and becoming. Salviati's position is that there should be no distinction between the two realms, the physics of the heavens being the same as the physics on Earth. This, of course, was supported by Galileo's famous astronomical observations that showed, for example, that the Moon and the Sun were far from perfect.

The second day is dedicated to a host of Aristotelian arguments against the motion of the Earth, such as the claim that objects such as clouds or falling balls would be left behind if the Earth rotated.

Here Galileo brings all his knowledge of the physics of motion to show that the objects would share in the Earth's motion; for an observer watching a ship sail by, a stone dropped from the ship's mast would not be left behind since it is released with the same horizontal speed as the ship. The stone's motion will thus consist of two combined motions, a forward motion with constant speed, and a downward motion with constant acceleration (free fall). This is an example of "projectile motion," which would occupy Galileo during the last years of his life.

However, in successfully refuting all of Simplicio's objections to the motion of the Earth, Salviati still hadn't proved that the Earth indeed moves. The tension was building up for the grand finale, scheduled for the last day of the debate. On the third day, the issue of the Copernican vs. the Ptolemaic arrangement of the cosmos is addressed. For this argument, Galileo adopted a simplistic version of the Copernican model, neglecting all its complications such as epicycles, or the fact that the Sun was not truly the center, to show the bewildered Simplicio how superior it was to Ptolemy's scheme. Perhaps because of his erroneous belief in the "circular inertia" of heavenly bodies, Galileo never accepted Kepler's elliptical orbits. He preferred a more general approach, as opposed to the tiresome details of planetary motion. But it could be argued that Galileo deliberately misrepresented the Copernican model to serve his own purposes.

Finally, on the fourth day, the time had come for the definitive proof of the motion of the Earth, based on the theory of the tides. It was also time to satisfy Urban's requests. Early in the day, Simplicio suggests that the tides may be explained by God's miraculous interference. A sarcastic Salviati responds that since

> *we must introduce a miracle* to achieve the ebbing and flowing of the oceans, let us make the Earth miraculously move with that motion by which the oceans are naturally moved. This operation will indeed be much simpler and more natural among things miraculous, as it is easier to make a globe turn around . . . than to make an immense bulk of water go back and forth. [my italics]

First, it is clear that Salviati is being forced to introduce a miracle, that he does not believe a miracle is necessary. But much more important, it is Galileo, and not the pope, who knows best how God operates. If He chooses to make a miracle occur, it had better be the simplest possible miracle. Giving motion to the entire Earth once was a much simpler miracle than having to act miraculously every day to promote the ebbing and flowing of the oceans. Surely God was smarter than that.

After Salviati's detailed exposition of the theory of the tides, Simplicio brings in God's infinite power:

> Keeping always before my mind's eye a most solid doctrine that I once heard *from a most eminent and learned person, and before which one must fall silent,* I know that if asked whether God in His infinite power and wisdom could have conferred upon the watery element its observed reciprocating motion using some other means than moving its containing vessels, both of you would reply that He could have, and that He would have known how to do this in many ways which are unthinkable to our minds. From this I forthwith conclude that, this being so, it would be excessive boldness for anyone to limit and restrict Divine power and wisdom to some particular fancy of his own. [my italics]

Thus, Urban's words come from Simplicio's lips, the Aristotelian fool whom Salviati continuously humiliated during the debate.

In May 1630, Galileo went to Rome to make sure he could proceed with the publication of the manuscript. The pope received him for a long audience and confirmed that he had no objections to presenting the merits of the Copernican model, as long as it was treated as a hypothesis. He did complain about the title with its direct mention of the tides, and suggested *Dialogue on the Two Great World Systems* instead. Too busy to read the manuscript, he left the details of possible revisions to Father Niccolo Riccardi, Master of the Palace and Chief Censor and Licenser. Riccardi procrastinated. To his difficulties in dealing with the text, there was the added pressure by Galileo and his allies to get things going as fast as possible. There followed a clever display of political maneuvering by Galileo, which culminated with Riccardi consenting that the revision be done by a Florentine Inquisitor picked by Galileo himself. Still, fearing the worst, Riccardi sent two letters to Galileo. In the first letter, dated May 24, 1631, Riccardi reminded Galileo

> that it is the intention of Our Lord's Holiness that the title and subject should not be on the flux and reflux but absolutely on the mathematical consideration of the Copernican position concerning the motion of the Earth, so as to prove that, except for the revelation of God and the Holy Doctrine, it would be possible to save the appearances with this position, solving all the contrary arguments that experience and the Peripatetic [Aristotelian] philosophy could advance, so that the absolute truth should never be conceded to this opinion, but only the hypothetical, and without Scripture.

The book ended up being simply called *Dialogue.** Following the pope's original request, Riccardi granted Galileo the right to destroy the arguments of the Aristotelians in favor of Copernicanism, as long as it remained a "mathematical consideration," "hypothetical, and without Scripture." The absolute truth is reserved for Divine revelation and the Holy Doctrine. In the second letter, dated July 19, Riccardi reminds Galileo that he must "add the reasons from divine omnipotence dictated to him by His Holiness, which must quiet the intellect, even if it were impossible to get away from the Pythagorean [Copernican] doctrine." It seems that Riccardi was merely checking his own back, as if he could sense the winds gathering for the upcoming storm.

The first copies of the *Dialogue* came out in February 1632. By August the sale of the book was forbidden, and in October an almost seventy-year-old Galileo was summoned to appear once more before the Inquisition in Rome. According to Francesco Niccolini, then Tuscan ambassador in Rome, the pope was furious with Galileo. How could he have dared to subject God to necessity? He felt outwitted, deceived, and betrayed by someone he held very

---

*It carried a subtitle, in which Galileo clarifies that the book summarized a four-day meeting concerning the two chief systems of the world, Ptolemaic and Copernican. Only after 1744 was it entitled *Dialogue Concerning the Two Chief World Systems, Ptolemaic and Copernican.*

dear. To this personal outrage one must add the serious implications of Galileo's work during a time when Europe was divided by the Thirty Years War. This was no time to call into question the authority of the pope, and of the whole Catholic Church. Galileo had to be punished, and the moral integrity of the Church had to be maintained.

For reasons of poor health, Galileo delayed his arrival in Rome until February 1633. His first interrogation took place on April 12. The Inquisitor read to him the injunction of 1616, which stated that Galileo should not hold, teach, or defend the Copernican ideas in any way. To this Galileo replied that he did not remember the words 'teach' or 'in any way,' and provided a copy of Bellarmine's certificate, which indeed did not use these words. When the Inquisitor asked Galileo if he had shown Father Riccardi the injunction during the negotiations for the release of the *Dialogue*, Galileo replied that he did not think that was necessary,

> for I have neither maintained nor defended in that book the opinion that the Earth moves and that the Sun is stationary, but have rather *demonstrated the opposite of the Copernican opinion, and shown that the arguments of Copernicus are weak and not conclusive.*

I can almost see the mouths of the Inquisitors dropping open at such a statement. Was Galileo blinded enough by his mission that he thought his opponents were such fools? Amazing as this seems, this was the gist of his whole defense, that he didn't support Copernicus at all. Of course, this strategy didn't work.

A few days after the questioning, Galileo was approached unexpectedly by the Commissary of the Inquisition, Father Vincenzo Maculano. The result of the meeting is best summarized in a letter dated April 28, from Maculano to Cardinal Francesco Barberini, the pope's brother, who was one of the judges in the trial. Below I quote parts of the letter, which reveal quite clearly the Church's intentions:

> Their Eminences [the Inquisitors] approved of what has been done thus far and took into consideration, on the other hand, various difficulties with regard to the manner of pursuing the case and of bringing it to an end. More especially as Galileo has in his examination

denied what is plainly evident from the book written by him, since in consequence of this denial there would result the necessity of greater rigor of procedure and less regard to the other considerations belonging to this business. Finally I suggested a course, namely, that the Holy Congregation should grant me permission to treat extrajudicially with Galileo, in order to render him sensible of his error and bring him, if he recognizes it, to a confession of the same. . . . That no time might be lost, I entered into discourse with Galileo yesterday afternoon, and after many and many arguments and rejoinders had passed between us, by God's grace, I attained my object, for I brought him to a full sense of his error, so that he clearly recognized that he had erred and had gone too far in his book. . . . I trust that your Eminence will be satisfied that in this way the affair is being brought to such a point that it may soon be settled without difficulty. The court will maintain its reputation; it will be possible to deal leniently with the culprit. . . . It will only remain to me further to question him with regard to his intention and to receive his defense plea; that done, he might have his house assigned to him as a prison, as hinted to me by your Eminence. . . .

In the next hearing, Galileo confessed to his errors. He even went as far as suggesting that he would write a follow-up to the *Dialogue*, adding one or two more days in which he promised "to resume the arguments in favor of the said opinion [the motion of the Earth and the stability of the Sun], which is false and has been condemned, and to confute them in such most effectual manner as by the blessing of God may be supplied to me."

Galileo was clearly afraid. He realized he could not bend the entire Church by the force of argument alone. He had no choice but to submit to the Inquisitors' demands. They achieved their goal of humiliating a man who thought himself invincible. Fortunately for Galileo, the Inquisitors dropped his offer of writing a continuation to the *Dialogue*, and after a few more hearings, they finally arrived at a sentence. The *Dialogue* was to be prohibited; Galileo had to abjure the Copernican opinion, was sentenced to house arrest until the end of his life, and for three years had to repeat once a week the seven penitential psalms. (It was arranged that Galileo's daughter, Maria Celeste, a Carmelite nun, would recite the psalms for him.)

And so it was that on June 22, 1633, kneeling in front of the Inquisitors, Galileo's voice echoed within the hollow walls of the convent of Santa Maria sopra Minerva,

> . . . with sincere heart and unfeigned faith I abjure, curse, and detest the aforesaid errors and heresies and generally every other error, heresy, and sect whatsoever contrary to the Holy Church, and I swear that in the future I will never again say or assert, verbally or in writing, anything that might furnish occasion for a similar suspicion.

According to a doubtful legend, as Galileo rose from his knees, he muttered the words *eppur si muove*, "and yet it moves." True or not, the legend symbolizes the steadfast belief he had in his ideas, which not even the humiliation of the long trial would shatter.

Certainly Galileo had the last word. Copies of the *Dialogue* were smuggled out of Italy, and in 1635 a Latin translation was widely available in Europe. When he returned home to Florence, he started to work on what would become perhaps his masterpiece, the *Dialogues Concerning the Two New Sciences*, published in Leyden in 1638.* In the book Galileo applied his principle that nature always acts in the simplest possible way to present a quantitative analysis of motion. Through a combination of experimentation and geometrical deduction, he obtained the mathematical relations describing the motion of falling bodies and projectiles, which would prove instrumental to the Newtonian synthesis of motion in the heavens and on Earth. At the third day of the discussions, Sagredo prophetically remarks,

> I really believe that just as, for instance, the few properties of the circle proven by Euclid in the Third Book of his *Elements* lead to many others more recondite, so the principles which are set forth in this little treatise will, when taken up by speculative minds, lead to many another remarkable result; and it is to be believed that it will be so on account of the nobility of the subject, which is superior to any other in nature.

*Guess who are the three characters in the book? And yes, Simplicio continues to be the dumb Aristotelian.

Galileo died in 1642, the year Isaac Newton was born. All but three of his bones rest in the Church of Santa Croce, next to the remains of Michelangelo and Machiavelli. The missing ones, those of the middle finger of his right hand, are displayed under a glass dome in the Museum for the History of Science in Florence, pointing defiantly at the passersby.

Galileo's episode with the Church serves as a powerful reminder of how excessive ambition may corrupt even the most sincere devotion. This is as true for Galileo as for the pope and the Holy Inquisitors. It is too easy simply to blame the Church for what happened, to say that Galileo's voice of reason and intellectual freedom was silenced by ignorance and the fear of change. It is true that the Church's action opened a schism between science and religion that is still very much alive today. The Church also failed to recognize that Galileo's voice was not that of a heretic, but that of a new worldview, a challenge not only for Catholics, but also Protestants, Jews, and Muslims. But Galileo's approach couldn't have been more disastrous. In his crusade against the prevailing ignorance, he failed to understand his own limitations and overlooked the power of his opponents. It is frustrating to think that maybe, with a little more tact, he could have achieved his goal, although history doesn't care much for this sort of speculation. Galileo's actions turned the very Church he so eagerly tried to serve against him, against his ideas, and against his followers. He did not suffer the pains of torture, nor was he imprisoned in a dungeon, but he had his most fundamental right removed, the right to freely express his thoughts and beliefs. But the Church had acted too late, clumsily trying to smother with blankets a fire of epic proportions.

5

THE TRIUMPH OF REASON

Nature and Nature's Laws lay hid in Night:
God said, Let Newton be! and All was Light.
— ALEXANDER POPE

How far can we go in exalting Newton's scientific achievements? Not far enough. Few minds in the intellectual history of humankind have left such an imprint as Newton's. His work represents the culmination of the Scientific Revolution, a grandiose solution to the problem of motion that had haunted philosophers since pre-Socratic times. In doing so, he laid the conceptual foundations that were to dominate not only physics but also our collective worldview until the dawn of the twentieth century.

Newton's tremendous impact on the intellectual development of Western culture can be traced to the effectiveness with which he applied mathematics to physics. With disarming clarity of thought, he showed that all motions observed in nature, from the familiar falling of a raindrop to the cosmic racing of comets, are understandable in terms of simple mathematical laws of motion. Quantitative reasoning became synonymous with science, transforming Newton's successful methodology into a blueprint for all areas of intellectual activity, not only scientific, but also political, historical, social, and moral. As remarked by the British philosopher and educator Isaiah Berlin, "No sphere of thought or life escaped the consequences of this cultural mutation."

Newton's genius knew no bounds. His appetite for learning far transcended what we would nowadays call science. He devoted an equal amount of time to studies in alchemy and theology, dealing with arcane questions that ranged from the transmutation of elements to biblical chronology and the nature of the Christian Trinity.

Although we correctly learn in school that Newtonian physics is a model of pure rationality, we would dishonor his memory if we overlooked the crucial role God plays in his universe. It may be true that when we look at Newton's scientific achievements we can neglect the more metaphysical side of his personality. But that is only half the story. For Newton saw the Universe as a manifestation of the infinite power of God. It is no exaggeration to say that his life was one long search for communion with the Divine Intelligence, which Newton believed endowed the Universe with the beauty and order manifest in nature. His science was a product of this belief, an expression of his rational mysticism, a bridge between the human and the Divine.

THE BLOSSOMING OF GENIUS

The story of this "sober, silent, thinking lad" starts on Christmas Day, 1642, in the manor house of Woolsthorpe, in Lincolnshire, England. There we find frail little Isaac, born early that day to Hannah Ayscough Newton. Newton's father, also called Isaac, had died three months earlier, leaving Hannah in charge of their property. The Newtons were part of a small minority that managed to prosper while an increasing concentration of land was deepening the social differences in rural England. But despite their mild affluence, the Newtons were uneducated. In fact, it was thanks to the Ayscoughs' influence that little Isaac was to become the first Newton who could sign his name. Hannah could write, though barely, as some—quite affectionate—letters from her to Newton show; and her brother William was an ordained Anglican minister in nearby Colsterworth, with a degree from Cambridge University.

Born fatherless, sickly Isaac was soon to become motherless. When he was three years old, Hannah married Barnabas Smith, a sixty-three-year-old minister, and moved to the nearby village of

North Witham. Smith did not want Isaac living with them in Witham and made sure he stayed with Grandma Ayscough in Woolsthorpe, which he gladly paid to have restored. His mother's departure left an emotional void that was to curse Newton for the rest of his life. He never married and, as far as we know, died a virgin. His feelings were to be locked inside, channeled, as it were, to an obsessive, furious devotion to his creative work.

An entry in a list of sins written by Newton nine years after his stepfather's death, in which he recalls "threatening my father and mother Smith to burn them and the house over them," sums up his feelings toward Barnabas and his estranged mother. Such anger and frustration were to mark him deeply for life. He grew up to be a tortured and bitter man, mistrustful of others, and, later in his life, always on the verge of a nervous breakdown.

When Newton worked on a problem, the world outside ceased to exist. He would forget to eat, drink, and sleep, only giving up to his body's cries of despair with reluctance. When I was a postdoctoral fellow at Fermi National Accelerator Laboratory, I was always amazed to find so many colleagues hard at work on a Sunday afternoon, or on a Saturday night. Of course I was there too, but still, we would all go downstairs at some point to get some (very bad) coffee or a chocolate bar, and eventually we did go home to catch some sleep. Newton just worked. While most scientists try to solve a problem by holding it in their minds for a few moments, dropping it, and trying to catch on later, Newton could hold a problem in his mind for hours and even weeks on end. He possessed astonishing powers of concentration, unequaled intuition, obsessive devotion, and a remarkable mathematical ability.

Newton's mother saw that he received an education, hoping that he would eventually help manage her property, especially after the death of Barnabas in 1653. But it soon became clear that he had no interest or talent for agricultural matters. He was not a particularly distinguished student either, at least within the established curriculum of a country school in those days, which consisted mostly of the rudiments of Greek, Latin, and some Hebrew. However, he did become acquainted with geometry through the notebooks of Henry Stokes, the headmaster of the small single-room King's School at

Grantham. Under Stokes's mentorship, Newton was accepted by Trinity College at Cambridge University in the spring of 1661.

Even though Copernicus had published *De revolutionibus* over a hundred years earlier, and Kepler and Galileo had long overthrown the foundations of the medieval worldview, the Cambridge curriculum was still steeped in Aristotelian thought. As in most European universities at that time, a liberal education consisted of the *trivium* (rhetoric, grammar, and logic), followed by the *quadrivium* (geometry, arithmetic, music, and astronomy), which included an introduction to Aristotelian physics along with Euclidean geometry. However, sometime during 1664 Newton's life took a decisive turn; he discovered the works of the French philosophers René Descartes and Pierre Gassendi, and others who were proposing a completely novel way of looking at the world. Newton devoured their books, compiling a list of queries in a blank notebook he got from Barnabas Smith's library. On the top of the notebook he wrote *"Amicus Plato amicus Aristoteles magis amica veritas,"* which roughly means: "Plato is my friend; Aristotle is my friend; but my best friend is truth." He had discovered his true vocation; he would become a natural philosopher.

Through the works of Descartes and Gassendi, Newton was introduced to the new "mechanical philosophy," a term coined by the great chemist Robert Boyle, with whom Newton would later share his interest in alchemy. They were the embodiment of a new attitude toward nature, expressed forcefully by Francis Bacon, who proclaimed a combination of deductive reasoning and experimentation as the only viable approach to "mastering" nature.*

Descartes made a clear distinction between mind and matter, mind being indivisible, the seat of the self (the "I" from the famous "I think, therefore I am" statement), while matter was infinitely divisible, an inert medium through which mind could operate. Matter had extension, while mind didn't. Every phenomenon in nature could be understood in terms of mechanical interactions between its material components. However, contrary to the atomistic view,

---

*There are several references in Bacon's writings to nature as the unruly female that must be conquered by reasoning, which he clearly assumed to be a male characteristic. This unfortunate metaphor may well have influenced the characterization of science as an exclusively male enterprise, as well as the irrational exploitation of nature during the Industrial Revolution, both of which survive to our days.

which held that indivisible atoms roamed through empty space, Descartes postulated that space cannot be empty, being filled with some kind of matter.

In order to explain the motions in the solar system, Descartes devised an intricate system of vortices that whirled the planets in orbits around the Sun, very much like corks floating around a draining sinkhole. Light was a kind of pressure propagating in this material "plenum," like a wave in the ocean. Descartes's arguments strictly deny the possibility that material bodies may influence each other without coming into physical contact. He ruled out "action at a distance" as a form of animism.

Against Descartes's dualistic mind/matter philosophy stood Gassendi's atomism, firmly rooted in the old atomistic tradition we encountered before. Gassendi postulated that light was not pressure through a medium, but atoms moving in a void with immense velocity. Newton's notebook is filled with discussions on Cartesian and atomistic ideas, although he clearly leaned toward atomism from early on.

By the time he obtained his B.A. in the spring of 1665, Newton had not only assimilated the works of his predecessors, but had also started an investigation of the physics of motion and of light that would shape the rest of his scientific career. I stress the word *scientific*, because that was not the only focus of his attention. Far from it. A proper appreciation of Newton's intellect cannot neglect his lifelong devotion to alchemy and theology. Newton was a multidimensional person who tried to understand the world around him in numerous ways. He attacked alchemy and theology in the same way he attacked natural philosophy; he read everything that had been written before on the subject, and proceeded to rewrite things his own way. The historian Betty Jo Teeter Dobbs, an expert in Newton's alchemical writings, asserts that Newton explored "the whole vast literature of the older alchemy as it has never been probed before or since." In his devotion to alchemy, Newton may have been searching for a spiritual quality absent in the rigors of his work in mechanics. This may also explain his equally intense devotion to theology, which Newton elaborated on in several million words, largely unpublished to this day.

Another possibility is that science, alchemy, and theology represented complementary aspects of Newton's search for the Divine. The fact that science is rational does not necessarily divorce it from the Divine. This separation is very much dependent on the scientist's subjective interpretation of what "Divine" means. We saw how the Pythagorean mystical interpretation of numbers as the language of nature was encompassed by Kepler's geometrical construction of the cosmos. The power of mathematics to describe the world around us never ceases to astonish. To Newton, nature was a manifestation of God's infinite intelligence. The rationality of his science was charged with spirituality.

THE APPLE OF KNOWLEDGE

Between the summer of 1665 and the summer of 1667, continuous outbreaks of the plague forced Newton to return from Cambridge to Woolsthorpe. During these two years his genius exploded with a force that is hard to fathom. The explosion did not come out of the blue, however. Newton's notes from 1664 and early 1665, mostly unknown to any of his colleagues or teachers, already show a mastery of mathematics probably unrivaled by anyone in Europe at the time. What he achieved during these two years was certainly built upon the solid foundations of his recently acquired knowledge. But the originality and sheer quantity of ideas is indeed astonishing. This is how Newton himself recalled, perhaps with some nostalgic liberty, these two years, which early biographers called the *anni mirabili,* or "marvelous years":

> In the beginning of the year 1665 I found the method of approximating series and the rule for reducing any power of any binomial into such a series. The same year in May I found the method of tangents of Gregory and Slusius, and in November had the direct method of fluxions [what we call differential calculus today*] and the next year in January had the theory of colors and in May following I had entrance into the inverse method of fluxions [integral calculus]. And the same year I began to think of gravity extending

*The calculus was independently invented by the German Gottfried Wilhelm Leibniz. Later in his life, Newton would unjustly accuse Leibniz of plagiarism.

to the orb of the Moon and . . . from Kepler's rule of the periodical times of the planets being in sesquialterate [three-halves power] proportion of their distance from the center of their orbs, I deduced that the forces which keep the planets in their orbs must be reciprocally as the squares of their distances from the centers about which they revolve; and thereby compared the force requisite to keep the Moon in her orb with the force of gravity at the surface of the Earth, and found them answer pretty nearly. All this was in the plague years 1665–1666. For in those days I was in the prime of my age for invention and minded mathematics and philosophy more than any time since.

Let us examine some of Newton's discoveries during the plague years, for they would play a crucial role in his later work in optics, mechanics, and gravitation. His mentor at Cambridge, Isaac Barrow, the first Lucasian Professor of Mathematics, got him interested in the study of optics. Newton experimented with prisms (a crystal shaped like a pyramid), lenses, and mirrors, trying to understand the physical properties of light. He knew that when sunlight passes through a prism, it refracts into the seven colors of the rainbow, red, orange, yellow, green, blue, indigo, and violet. Going beyond the prevailing style of the time, he made a series of *quantitative* studies of the nature of this refraction and measured the different angles by which different colors deviate from their original paths. His emphasis on experimentation, as opposed to qualitative descriptions based on sensory observation of the phenomena, was a crucial difference between Newton's approach and that of others.

From his delicate and accurate experiments, Newton realized that the reason why different colors refracted differently had to do with their velocities as they traversed the prism, which acted as a kind of color "filter"; the more slowly the color moved through the prism, the more it deviated from its original trajectory. Plus, by adjusting the angles of several prisms laid on top of each other, he could make different colors recombine into white light again. Newton concluded that white light was nothing more than the superimposed combination of the seven colors of the rainbow. This counterintuitive result was exactly opposite to the then accepted theory of colors, which stated that white light was pure, and that colors appeared as white light was modified by interacting with different media, such as a prism, or as it was reflected from the surface of a colored body. Newton believed that his results strongly supported a corpuscular, or atomistic, theory of light, each color being made of a different kind of atom, which remained unaltered as it propagated through different media. But he would not commit to this view until much later.

Newton was also unhappy with Descartes's explanation for the motion of the planets, based on his theory of cosmic vortices. To Newton, and to many other natural philosophers of the seventeenth century, the greatest challenge in the emerging new science of motion was to provide a proper explanation for the stability of the planetary orbits. That involved not only an understanding of the nature of the forces that kept the planets in orbit, but also the need for a new mathematics to describe how their positions changed over time. That is, they had to find the solution to a mathematical equation that describes planetary motion. This new mathematics is what we today call calculus, and what Newton called *fluxion*. Once in possession of his new tool, Newton could proceed to lay the foundations of the new science of motion. But that search would extend far beyond the plague years.

While in Woolsthorpe, Newton became absorbed by the problem of circular motion, which also occupied the mind of a contemporary scientist, the Dutchman Christian Huygens. Huygens had coined the term *centrifugal* force to explain the outward force felt by a body in circular motion. By analogy with a stone being whirled around in a circular orbit, Newton thought of circular motion as a state of equilibrium between the centrifugal force and the tension on the string. Later he would realize that this was not quite correct, but that circular motion was the result of a *centripetal* force, pushing the body toward the center of its orbit. For example, if we could turn off the gravitational force attracting the Moon to the Earth, the Moon would fly away in a straight line, tangent to its orbit, its natural inertial motion. It would not fly outward, under the influence of a centrifugal force pushing it in that direction. It is the centripetal force that makes the orbiting body deviate from the inertial motion it would have otherwise, forcing it to "fall" toward the center of its orbit.

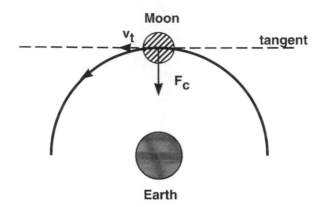

Circular motion: The centripetal force (F_c) points toward the center of the Earth, making an object "fall" toward the center of its orbit. If the Earth's gravitational attraction is "turned off," the Moon would fly away in the tangential direction with velocity v_t.

One day, tired of musing about the intricate properties of circular motion, Newton decided to go outside for a break under one of the nice apple trees at his mother's home. We all know the story: As he watched an apple fall to the ground (or was it on his head?), he wondered if the same force that brought the apple down could not also be acting on the Moon, keeping it in its orbit. He knew of Galileo's description of free fall as an accelerated motion downward. Maybe the Moon was also falling, but its fall was balanced by the centrifugal force, creating a circular motion as a result. Although Newton's reasoning was not quite correct, he could still use it to prove that the force acting on the Moon and on the apple decreased with the square of the distance to the Earth, at least "pretty nearly." This deduction is not yet his theory of universal gravitation of twenty years later, but the first seeds started to germinate. In the words of the historian of science Richard Westfall, "Some idea floated at the border of his consciousness, not yet fully formulated, not perfectly focused, but solid enough not to disappear. He was a young man. He had time to think on it as matters of great moment require." Great theories don't simply appear in someone's head as if by magic, but take time to blossom. The so-called "eureka" cry is more an exclamation of mental relief than a cry of sudden revelation.

The System of the World

Newton returned to Cambridge in April 1667, ready to move up on the academic ladder. In October he became a fellow of Trinity College, solemnly swearing that "I will embrace the true religion of Christ with all my soul . . . and will either set Theology as the object of my studies and will take holy orders when the time prescribed by these statutes arrives, or I will resign from the College." Despite his new status, Newton continued to be as secluded as ever, indifferent to social life. On the rare occasions in which he would engage in social contact, his awkwardness only served to further increase his self-imposed isolation. On one occasion Newton invited some friends to his chamber. As he went to his study to fetch more wine, some thought came into his head. So absorbed did he become that he completely forgot about the wine and his friends waiting downstairs. Either his mind was indeed very far away, or the company was extremely boring. Absentmindedness can be a very useful skill.

Eccentricities aside, Newton's genius was fully appreciated by the Cambridge community. During the summer of 1669, Isaac Barrow sent a copy of Newton's recent work on infinite series* to the mathematician John Collins, with a cover letter stating that the work was by "Mr. Newton, a fellow of our College and very young (being but the second youngest Master of Arts) but of an extraordinary genius and proficiency in these things." At that time Barrow was thinking about resigning his appointment as Lucasian professor in order to dedicate himself to what he believed was his true calling, theology. Barrow saw in Newton a more than worthy successor and played a large part in his appointment as the new Lucasian professor in October 1669. This professional jump launched Newton into the highest circles of Cambridge's academic hierarchy. Not bad for a man of twenty-six.

As the Lucasian chair came with teaching requirements, Newton dedicated his first series of lectures to the study of light, which included (and expanded) his results with prisms and lenses obtained

---

*Infinite series refer to an infinite sequence of numbers, obeying some rule, such as the series $1 + \frac{1}{2} + \frac{1}{4} + \frac{1}{6} + \frac{1}{8} + \ldots$.

earlier in Woolsthorpe. His lecture notes would eventually appear in his last masterpiece, *Opticks,* published in 1704. He also found time to set up a laboratory where he could pursue his studies in alchemy, which would occupy much of his time for the next twenty years. Both of these activities, different as they were in method and goals, required manual dexterity and mechanical skills which New-ton also possessed. Experimenting with mirrors and lenses, which he ground using tools of his own design, Newton developed a new kind of telescope, a *reflector,* which produced images of much better quality than the *refractor* telescopes in use at the time.

For a change, Newton didn't keep this invention quiet. News about this wonderful new instrument quickly reached the ears of members of the Royal Society in London, an institution devoted to the development of the "new science," as dictated by the philosophy of Bacon and Descartes.

Newton's telescope caused a sensation, just as Galileo's had sixty-one years before. For the second time in the history of science, the telescope launched a scientific career into celebrity status. In January 1672, a month after Barrow took the new telescope to Lon-don, Newton was elected Fellow of the Royal Society. Excited by his newfound fame, he quickly put together a manuscript on his theory of colors, mailing it to London in February. He was soon to regret his hastiness.

The Royal Society was not Cambridge, where Newton's intellect reigned supreme. Other able minds were at work in England at the time, and Newton had to face their scrutiny. The paper on optics was severely (and wrongly) criticized by Robert Hooke, who claimed to have done the same experiments as Newton and obtained oppo-site results. Hooke considered himself the authority on the subject and didn't welcome a challenge to his status. Also critical was Chris-tian Huygens, who was proposing a wave theory of light, as opposed to the corpuscular theory embraced by Newton.* Hooke and Huy-gens clashed with Newton in a series of letters that became progres-sively less polite. It is in one of these letters to Hooke that we find Newton's famous remark, "If I have seen further [than you] it is by

---

*As we will see when we discuss quantum mechanics, both Newton and Huygens were, in a loose sense, right and wrong, although for reasons they could never have guessed.

standing on the shoulders of Giants." Out of context, Newton's statement seems to have been uttered in pure modesty, when in fact, it was a sarcastic remark directed at Hooke's smaller intellectual (and physical, as Hooke was nearly a dwarf) stature.

As a result of these disputes, Newton isolated himself once again within the walls of Trinity College. As he wrote to Henry Oldenburg, the secretary of the Royal Society, all these "frequent interruptions that immediately arose from the letters of various persons (full of objections and of other matters) . . . caused me to accuse myself of imprudence, because, in hunting for a shadow hitherto, I had sacrificed my peace, a matter of real substance." He was burning with desire to pursue other matters, which included his alchemical and theological explorations, and from 1678 on, he avoided most disputes concerning his scientific work, leaving them to be fought by his friends. Physics and mathematics would be kept somewhat on the back burner until August 1684, when Newton received a visit from Edmond Halley, also a member of the Royal Society, popularly known for tracing the orbit of the comet that bears his name.

Halley came to Newton because he was stuck on a physics problem. Together with Hooke and Christopher Wren, famous for designing St. Paul's Cathedral in London, Halley had shown that the Sun must exert a force, varying with the inverse square of the distance, which keeps the planets in orbit. This they deduced from Huygens's work on circular motion and Kepler's third law, just as Newton had twenty years before. However, the question remained as to what sort of orbits the planets would trace under the influence of such a force. When the three men met at the Royal Society in January, Hooke claimed he knew the answer, but would keep it secret until others realized how hard the problem was. He was, of course, bluffing, probably buying time so that he could work out the answer. To add spice, Wren offered an award to the first person who could provide the answer, a book worth forty shillings. Not much compared to the immortal glory of solving an enigma of such proportions, Wren claimed.

This is how Newton recalled the meeting with Halley, as retold many years later by the mathematician Abraham DeMoivre:

In 1684 Dr. Halley came to visit him at Cambridge; after they had been some time together, the Doctor asked him what he thought

the Curve would be that would be described by the Planets suppos-
ing the force of attraction towards the Sun to be reciprocal to the
square of their distance from it. Sir Isaac replied immediately that
it would be an Ellipsis; the Doctor struck with joy and amazement
asked him how he knew it, why saith he, I have calculated it, where-
upon Dr. Halley asked him for his calculation without any farther
delay; Sir Isaac looked among his papers but could not find it, but
he promised him to renew it, and then send it him. . . .

Newton knew very well where his papers were. But he had grown
wiser after the many disputes of the last twenty years and decided to
verify his results before making his discoveries public. In November,
Halley received a nine-page treatise entitled *De motu corporum in
gyrum*, "On the Motion of Bodies in an Orbit," usually called *De motu*
for short. It contained not just the answer to the original question,
but also the demonstration of Kepler's three laws from basic principles. This was much more than Halley could have hoped for. He
quickly understood that those nine pages represented nothing less
than a revolution in celestial mechanics.

Newton realized that a solution to the problem of orbital motion entailed a completely original formulation of the science of
mechanics, which was only hinted at in *De motu*. He had the mathematical tools, but he needed new physical concepts as well. Newton
applied himself to this task with a devotion that even for his standards was extreme. From August 1684 to the spring of 1686, he dis-
appeared from social life. The exceptions were a few letters he
exchanged in early 1685 with John Flaamsteed, the first royal astronomer, asking for data on the comet of 1680–81, and on the motions of Jupiter's and Saturn's satellites. He later asked for data on
their relative motion when they are in conjunction (closest to each
other), and even on Flaamsteed's observations of the tides in the es-
tuary of the Thames. These questions reveal that he was already well
on his way toward formulating his theory of gravity.

Newton's masterpiece, the *Philosophiae naturalis principia mathe-
matica*, "Mathematical Principles of Natural Philosophy," known as
the *Principia* for short, was published in July 1687. No other work in
the history of science has been as fundamental in determining the
way we understand the physical world around us. Newton not only

developed a new science of motion based on the action of forces on material bodies, but he also showed that this science of motion is common to all bodies, be they on the Earth or in the skies. Using a rigorous mathematical approach, he joined physics and astronomy forever. All motions can be understood under the same unified framework independent of where they take place. There is only one physics, and its domain extends to the stars.

The *Principia* is built upon several revolutionary physical concepts, which took Newton many years to grasp. To truly appreciate Newton's accomplishment, and to set the stage for things to come, we should spend some time discussing a few of these concepts. The book opens with a set of definitions, terms Newton needed to formulate his mechanics. First, he introduces the concept of mass, which, believe it or not, was not properly defined until then. The mass of a body is what we often erroneously call its weight. It is a measure of the quantity of matter in a body. *Weight*, on the other hand, is how a certain mass responds to the acceleration, or pull, caused by gravity.* Thus, even though your mass is the same here or on the Moon, your weight will be different in the two places, since the acceleration caused by gravity is different on the Earth and on the Moon.

Newton then goes on to define the quantity of motion, or, as we call it today, the *momentum* of a body. Based on previous work by Galileo and Descartes, Newton defines the momentum of a body as the product of the body's velocity with its mass. Thus, a Volkswagen beetle and a Mack truck moving at twenty miles an hour have very different momenta, since their masses are so different. If you had no choice but to hit one of the two head-on, you know which one you would choose. Closely related is the concept of *inertia*, which can be understood as the reaction a body offers to any change in its momentum. Again, you know what inertia is from common experience: Moving a huge granite boulder from rest is much harder than moving a small one. Or, in Newton's new language, both the huge and small boulders have zero momentum initially, since their

---

*Even if you are not falling, gravity is accelerating you downward all the time. Just think of what would happen if suddenly the ground were removed from under your feet!

velocity is zero (they are at rest). However, because their masses are so different, the huge boulder offers much more resistance to changing its momentum (imparting some velocity to it) than the small one.

Now that the concepts of mass, momentum, and inertia are well established, Newton introduces the concept of *force*. It is the action exerted upon a body in order to change its momentum. You have to push on the boulders to make them move. That is, you have to apply a force on them to change their state of motion.

What Newton realized was that there were two ways by which you could change a body's state of motion: by changing the *magnitude* of its momentum or by changing the *direction* of its momentum. Again, a car is an excellent laboratory for explaining Newtonian physics. Imagine you are traveling on a straight road at forty miles an hour. As you accelerate to a higher speed, you change the momentum of the system (you and the car are the system). You feel the pressure pushing you back, the result of the force applied forward by the engine. But you are still on a straight road, and all you did was change the *magnitude* of the momentum, by increasing the car's speed. Now imagine turning on a curve but keeping the speed constant. The magnitude of the momentum is the same, but its *direction* changed as you turned on the curve. Only the action of a force can do that. You feel the centrifugal force pulling you outward while the tires keep you on the road. A pedestrian standing by the corner, who had never been in a car before, would reason that there was a force pulling you toward the center, forcing you to deviate from the straight-line motion. This force is what Newton called the *centripetal force*, to emphasize that it points toward the center.

Of course, all this talk about motion calls for a definition of how we measure it. Since motion has to do with change of position (relative to something fixed) in some period of time, Newton had to define what he meant by space and time: *Absolute space* is basically the geometric arena where physical phenomena take place. It remains indifferent to whatever is happening within it, a sort of inert container where bodies interact with each other according to well-defined physical laws. *Absolute time* flows resolutely forward, always at the same pace, impervious to the ways humans choose to keep it. That settled, Newton formulates his famous three laws of motion, which beautifully summarize all that we need to know when studying the motions of material bodies. In his own words:

> **LAW I:** Every body continues in its state of rest, or of uniform motion in a right line, unless it is compelled to change that state by forces impressed upon it.

"Projectiles continue in their motions, so far as they are not retarded by the resistance of the air, or impelled downwards by the force of gravity." Based on the discussion above, we know that the first law is an expression of the principle of inertia.

> **LAW II:** The change of motion is proportional to the motive force impressed; and is made in the direction of the right line in which that force is impressed.

Thus, change in motion, that is, change in momentum, is proportional to the force applied on the body. Change here may include both change in the magnitude or in the direction of the body's momentum. If the mass of the body doesn't change while a force is applied on it (as opposed to, say, a leaking water bucket being whirled around in a circle), then this law can be expressed as the famous $F = ma$ equation. F is the force, m is the mass of the body upon which the force is applied, and a is the acceleration that results from the application of the force: Thus, this law states, the change in momentum is due to a change in the velocity of the body, i.e., its acceleration.

> **LAW III:** To every action there is always opposed an equal reaction: or, the mutual actions of two bodies upon each other are always equal, and directed to contrary parts.

For example, a bow must be arched backward to propel an arrow forward. You can also experience this law most vividly by kicking a concrete wall.

There you have it, the foundations of the new science of mechanics, all appearing in the introductory pages of the *Principia*. The conceptual foundations made explicit, Newton finally started the volume itself, which he divided into three parts that he called books. The reading is complicated, for the books were written in thick geometrical language that made implicit use of his new mathematical tool, the calculus. It's sufficient to outline what's in each of the three books.

In Book I, "The Motion of Bodies," Newton applied his mechanics to the problem of motion under a centripetal force, showing what sort of orbits are possible, including, of course, circular and elliptical orbits. He then studied pendular motion, and the motion of bodies on curved surfaces, like a marble rolling inside a hollow spherical cavity. He concluded by proving that in treating the problem of a particle being attracted by a large spherical body, such as an apple attracted by the Earth, we can consider the large spherical body as a small body with all its mass concentrated at its center. This is a crucial step toward implementing the law of universal gravitation. In Book II, Newton examined the motion of bodies in the presence of friction, such as particles moving in a fluid. His main goal was to show that Descartes's vortices and "plenum" were not consistent with the stability of planetary orbits. On the basis of his calculations he concluded that interplanetary space is empty.

Book III Newton called "System of the World." Here we find the application of all the physics developed in the previous books to the problem of gravitational attraction. Newton's brilliant mind recognized that Galileo's description of projectile motion and Kepler's theory of elliptical orbits were in essence the same thing. This was no small feat. Kepler's laws were thought to deal exclusively with planetary orbits. Galileo's results on projectile motion were thought to deal exclusively with terrestrial motion. Heavenly motions and earthly motions were completely unrelated phenomena, governed by distinct laws. To complicate matters even more, Galileo had invoked circular inertia to deal with planetary motions, while Kepler put forth magnetism to explain the mysterious force pulling the planets in their orbits around the Sun.

Newton's breakthrough was to treat the Moon's orbit as a projectile motion. This is, quite schematically, how he reasoned: Imagine a cannon on the top of a very high mountain, as in the figure opposite. The trajectory of the cannonball is determined by how much initial velocity the cannon can impart to it. In the absence of gravity or friction, the cannonball would just continue in a straight line with constant velocity forever, as dictated by the principle of inertia. But gravity, being a centripetal force, deflects the cannonball's trajectory, making it fall with vertical acceleration. For small initial

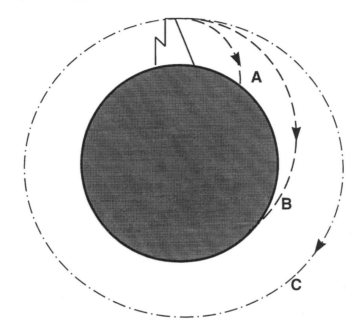

According to Newton's description of orbits as projectile motion, if a powerful cannon (too small to be seen in the picture) placed on top of a very high mountain shot a projectile with enough horizontal velocity, the projectile would "keep falling" as in path C, never hitting the ground. The combination of downward acceleration and large horizontal velocity will cause the projectile to orbit the Earth. For paths A and B, the horizontal velocity is not large enough, and the projectile falls to the ground.

velocity, the cannonball will drop close to the mountain's base. But we can imagine that, as the cannon's power is steadily increased, eventually the cannonball would have enough horizontal velocity to just "keep falling" toward the center of the Earth. The combination of the downward acceleration and horizontal inertial motion will cause the cannonball to circle around the Earth. The cannonball becomes a satellite in orbit! Of course, in real life we don't have a mountain that high or a cannon that powerful, and we have to propel our satellites upward with powerful rockets and then give them the horizontal velocity needed to set them in orbit. But the basic physics involved is essentially what Newton discovered over three hundred years ago.

Newton then completes the picture by showing that all material bodies attract each other gravitationally, with a force proportional to their masses and inversely proportional to the square of their distance. For example, two apples a foot away attract each other with a force four times stronger than they would two feet away. Thus, the Earth and the Moon attract each other, while the Sun attracts every object in the solar system and is equally attracted by all of them.

In addition to explaining all the motions in the solar system, including planets and comets, and even the precession of the equinoxes, Newton applied his theory to the phenomenon of the tides, showing that they result from the combined gravitational attraction of the Sun and the Moon on the oceans, burying forever Galileo's explanation based on the Earth's motions. In Newton's view, gravity was the universal glue, governing all motions on the cosmic scale. The system of the world was now an open book, unveiled in all its breathtaking beauty by the genius of the "sober, silent, thinking lad" from Woolsthorpe.

NEWTON'S GOD

"It may be justly said, that so many and so Valuable *Philosophical Truths*, as herein discovered and put past Dispute, were never yet owing to the Capacity and Industry of any one Man." Thus Halley concluded his review of the *Principia*. The work justly immortalized Newton as one of the supreme intellects in the history of humankind. It set the standards of how scientific works should be written and how scientific research should be performed. "We are to admit no more causes of natural things than such as are both true and sufficient to explain their appearances," Newton wrote in the *Principia*, stressing that "to this purpose the philosophers say that Nature does nothing in vain, and more is in vain when less will serve; for Nature is pleased with simplicity, and affects not the pomp of superfluous causes."

By writing these lines, Newton was not only echoing Galileo's emphasis on the simplicity of Nature, but also making sure that his approach, although incomplete, was sufficient to explain the phenomena involved. The incompleteness came from his inability to explain his assumption of gravity as *action at a distance*. The idea that gravity should act on all bodies from far away, without direct physical contact, was a hard one to swallow. It had led Kepler to associate gravity with magnetism, Galileo to introduce his "circular inertia," and Descartes to propose his theory of cosmic vortices. Newton showed that gravity had nothing to do with magnetism, and that circular inertia and cosmic vortices were artificial and erroneous constructs. But he could not justify his own hypothesis of action at a distance. It worked, and that was enough. As he wrote in the General Scholium to the *Principia* (a sort of concluding remarks):

> Hitherto we have explained the phenomena of the heavens and of our sea by the power of gravity, but have not yet assigned the cause of this power. This is certain, that it must proceed from a cause that penetrates to the very centers of the Sun and planets, without suffering the least diminution of its force. . . . But hitherto I have not been able to discover the cause of those properties of gravity from phenomena, *and I feign no hypothesis; for whatever is not deduced from the phenomena is to be called an hypothesis; and hypotheses, whether metaphysical or physical, whether occult qualities or mechanical, have no place in experimental philosophy. In this philosophy particular propositions are inferred from the phenomena, and afterwards rendered general by induction.* [my italics]

This is, to date, the credo of science. It is what distinguishes science from any other human endeavor. Science makes sense only if rooted in a rigorous methodology, built upon the interplay between experimentation and deduction.* A hypothesis that cannot be put to the test by quantitative experimentation must remain a hypothesis. It cannot belong to science, at least in the most rigorous definition adopted by Newton and his followers. In principle (and I stress the *in*

---

*Of course, this discussion leaves aside pure mathematics, which operates under a different set of rules.

principle), there should be no room for subjectivity of interpretation in science, although research at the cutting edge is often interpreted in different ways before conflicting opinions coalesce. Otherwise, science would lose its universality. Subjectivity may appear in the scientific creative process, but not in its final product. Thus, although Newton couldn't understand the cause of gravity's power, he opted to leave the question aside, "feigning no hypothesis."

This did not mean that Newton believed that "action at a distance" was possible. Five years after the publication of the *Principia*, he exchanged a few letters with Richard Bentley, the chaplain to the Bishop of Worcester and a brilliant theologian and classical scholar. Bentley was preparing the first set of Boyle Lectures on the theme "the observed structure of the Universe could only have arisen under God's guiding hand." As the lectures were to be the first popularization of Newton's ideas, Bentley asked Newton for his advice on several topics, among them the nature of gravity and the infinitude of the Universe. On the nature of gravity Newton wrote:

'Tis unconceivable that inanimate brute matter should (without the mediation of something else which is not material) operate upon and affect other matter without mutual contact; as it must if gravitation in the sense of Epicurus [atomism] be essential and inherent in it. And this is one reason why I desired you would not ascribe innate gravity to me. That gravity should be innate inherent and essential to matter so that one body may act upon another at a distance through a vacuum without the mediation of any thing else by and through which their action or force may be conveyed from one another is to me so great an absurdity that I believe no man who has in philosophical matters any competent faculty of thinking can ever fall into it. Gravity must be caused by an agent acting constantly according to certain laws, but whether this agent be material or immaterial is a question I have left to the consideration of my readers.

Does this confession of the limitations of his theory of gravity weaken its predictive power? Not at all. This is an issue that scientists have to deal with all the time. Contrary to what some people think, scientists don't have all the answers. What we can offer are testable principles that describe a large variety of phenomena. But if pressed hard, we must stop at some point and say, with Newton, that we feign no hypothesis. This is not a weakness of science, but simply the way it is built. We always strive to understand the physical causes behind observed (and predicted) phenomena as completely as possible. Every theory has its own limitations, and through the conceptual gaps left open by old theories, new theories emerge. This, in a nutshell, is how science perpetuates itself. For example, Newtonian mechanics cannot be used for very fast motions, say, with speeds approaching the speed of light. For that we need Einstein's theory of relativity. But for the low velocities of our everyday life, Newtonian mechanics remains "the" theory.

The *Principia* can be read as a book based on strict mechanistic principles, with no room for any metaphysical speculation. This is how we learn Newtonian physics in school. And this is perfectly fine, as long as we are only interested in the physics, and not in the cultural, historical, or psychological context within which the physics was conceived. Because of the way science is structured, it lends itself to this purely "operational" approach. This serves the practice of science well, but perhaps not its public understanding. For it is precisely this way of approaching science that renders it "cold" in the eyes of those less inclined to appreciate the technical beauty of the subject. There is more to science than its technical language. The beauty of science lies in its power to bring us closer to nature. Of course, this is also its downfall, for an understanding of nature's ways implies a means of exploiting it.

With this said, let us go back to Newton's General Scholium, where the author makes manifest his veneration for the beauty of nature, which he presents as evidence for a Cosmic Designer:

> This most beautiful system of Sun, planets, and comets, could only proceed from the counsel and dominion of an intelligent and powerful Being. . . . We know him only by his most wise and excellent contrivances of things, and final causes; we admire him for his perfections; but we reverence and adore him on account of his dominion: for we adore him as his servants; and a god without dominion, providence, and final causes, is nothing else but Fate and Nature.

Blind metaphysical necessity, which is certainly the same always and everywhere, could produce no variety of things. All that diversity of natural things which we find suited to different times and places could arise from nothing but the ideas and will of a Being necessarily existing.

In Newton's correspondence with Bentley we can find his strongest arguments for a Cosmic Designer and for the infinitude of the Universe. Newton equated God with a Cosmic Geometer, the intelligent First Cause of all motions in the Universe:

To make this system therefore with all its motions, required a Cause which understood and compared together the quantities of matter in the several bodies of the Sun and planets and the gravitating powers resulting thence. . . . And to compare and adjust all such things together in so great a variety of bodies argues that cause to be not blind and fortuitous, but very well skilled in Mechanics and Geometry.

Bentley then asks Newton how a finite spherical Universe, operating under the force of gravity, could avoid collapsing into a giant mass located at its center. Newton replies that only in an infinite Universe could an infinite number of bodies remain in equilibrium. Since any body will be surrounded by an infinite number of bodies in all directions, the sum total of the gravitational attractions will vanish and the body will be static. He conceded that such a hypothesis was quite implausible, comparing it to balancing an infinite number of needles upon their points. "Yet," he writes, "I grant it possible, at least by divine power; and if they were once so placed I agree with you that they would continue in that posture without motion forever, unless put into new motion by the same power." Thus, God acts continuously to keep the Universe in check, either by rendering it stable or by causing the observed motions of the heavenly bodies. An infinite Universe, with all its complexities, was living proof of the existence of an omniscient and omnipresent God.

Newton's life after the publication of the *Principia* took a drastic turn, propelled from the academic world into the highest circles of

London's society. He was quite content to enjoy the unprecedented fame bestowed upon him. He became a member of Parlia-
ment in 1689, although it is said that he spoke only once, to ask
an attendant to close a window. He was not reelected. In 1696,
Newton became a warden of the Mint, and three years later he
assumed the post of master. He took immense pleasure in inter-
viewing counterfeiters, who trembled at the sight of those dark,
piercing eyes and stern face. He took his job seriously (wouldn't
he?), spending much energy implementing the recoinage scheme
of Lord Halifax. In 1703 Newton was elected president of the
Royal Society, a post he kept until his death, and Queen Anne
knighted him in 1705, an honor never before conferred on a scientist. He died in 1727 and was buried in the Jerusalem Chamber of
Westminster Abbey.

Amidst all this pomp, Newton still found time to supervise two
subsequent editions of the *Principia*. After Hooke's death in 1703,
he finally decided to compile his old results on the nature of light
in his other great work, the *Opticks*, which appeared in 1704. He
pressed on with his atomistic theory of light, stressing the divine origin of the fundamental building blocks of the Universe: "It seems
probable to me that God in the beginning formed matter in solid,
massy, hard, impenetrable, movable particles . . . even so very hard
as never to wear or break in pieces; no ordinary power being able to
divide what God himself made one in the first creation." A rational
God permeates Newton's work. In his infinite universe, reason was
the only possible bridge to the Divine.

From the mythic universes of our ancestors to the theo-
scientific musings of Newton, a common theme emerges: a deep
association of nature with the Divine, fueled by an overwhelming
desire to understand the Universe and our place in it. The rich ta-
pestry of ideas that we have covered so far is interwoven by this
common theme that is at the very root of early Western science.
It is therefore somewhat surprising that, nowadays, when one
talks about science, there is usually no mention of religion, unless
to stress that the connection between the two should not be
made.

There are several reasons why this is so, but perhaps the most relevant here is the concern scientists have with the legitimacy of scientific thought. Newton's tremendously successful rational approach to physics became an emblem of a new age for humankind, based on the power of thought as opposed to the power of faith. For centuries Europe had been plagued by the endless religious conflicts of the Reformation and Counter-Reformation. Times were ripe for radical change. If science could be formulated in a purely rational fashion, then it could potentially become the liberating voice of a new era, in which religious differences and dogmatism would be replaced by universal values.

In its purest version, this extreme rationalism was believed powerful enough to explain all natural phenomena without invoking God's presence; God should be banned from the picture altogether. As we will see, these ideas form the core of the Enlightenment movement of the eighteenth century, which attempted to implement this separation between science and religion. Of course, the degree to which this separation really occurred varied in practice, ranging from the strict atheism of Pierre-Simon de Laplace to the rational Christianity of Benjamin Franklin. But as science became more and more efficient in explaining an ever-increasing number of phenomena, the belief in this separation grew stronger, to the point of complete independence: The official scientific discourse did not tolerate any mention of religion. The role of religion in science went through a radical shift, from an explicit player to a "forbidden," almost embarrassing, memory.

How necessary is this separation between science and religion? Very. It serves as a protective device against subjectivism in the practice of science, insuring that it remains a common language in a very diverse community. The scientific discourse is, and should be, devoid of any theological content. Invoking religion to fill in the gaps of our scientific understanding today is, in my view, an anti-scientific attitude. If there are gaps in our knowledge (and there are plenty of them!), we should try to fill them with more science and not with theological speculation. In other words, it is not the "God of the gaps" approach that brings religion into science. If we want to find a place for religion in modern science we must look at the subjective motivations of individual scientists, and not at the final product of their scientific work.

In assuming this position, I am siding with Einstein, who would write that "religion without science is blind, and science without religion is lame." By this he meant that when it comes to natural phenomena, religion should not close her eyes to scientific advancements, as Galileo's trial painfully reminds us. But perhaps more surprisingly, Einstein believed that science needs a sort of religious inspiration to go on. Or, even more dramatically, that the devotion to science and the implicit faith we have in the power of human reason to pierce through the mysteries of nature are essentially religious. We won't (and shouldn't!) find the word *God* or *religion* in a scientific paper. But in many cases, an essentially religious component acts as a fuel for scientific research today, just as it did, in a more transparent way, for Kepler and Newton. It's just a matter of how broadly we define religion.

With this in mind, we now continue our journey toward twentieth-century cosmology, by describing the successes and limitations of the classical worldview.

PART III

○

THE CLASSICAL ERA

6

THE WORLD IS AN INTRICATE MACHINE

Napoleon: Monsieur Laplace, why wasn't the Creator mentioned in your book *Celestial Mechanics*?
Laplace: Sir, I have no need for that hypothesis.

—ANONYMOUS ACCOUNT

The ground-breaking scientific achievements of Kepler, Galileo, Descartes, Newton, and many others during the seventeenth century triggered a deep revision of the Western conception of the cosmos. Gone was the walled-in Universe of the Dark Ages, shattered by Newton's arguments for an infinite Universe as the only possible realm of an infinitely powerful God. Gone was the power of religious dogmatism to dictate scientific truth. And gone were the days when pure scholastic speculation could substitute for a science based on the interplay between experimentation and theory.

The rational foundation for the emerging new science built during the seventeenth century was expanded into a magnificent structure during the eighteenth. The physical world was reduced to pointlike masses moving under the influences of forces, as dictated by Newton's three laws of motion and by his law of universal gravitation. Built within this mechanistic approach to nature was a strict determinism: If we knew the positions and velocities of the objects in a given system (say, the Sun, the Earth, and the Moon) at a certain moment in time, using Newton's laws we could, *in principle*, predict the positions of the objects at any moment in time, both past and future! By the late eighteenth century, Pierre-Simon de Laplace (1729–1827) was very successful in explaining most motions in the solar system, while other Frenchmen such as Pierre-Louis Moreau de Maupertuis (1698–1759) and Joseph-Louis de Lagrange (1736–1813) had distilled Newtonian mechanics into powerful new mathematical formulism, rendering it applicable to a much wider range of complicated systems. The Universe was reduced to a giant clockwork mechanism, an intricate, but understandable, machine.

The inflated confidence in this strict determinism can be best illustrated by the belief held by Laplace and others that if a "supermind" could know the positions and velocities of all entities in the Universe at a given time, then it could predict the future of all things forever. Every movement, every thought, every surprise, good or bad, would be known to this giant intelligence. Destiny was perfectly predictable, the consequence of rigid mechanical laws. In this machinelike world there was no room for free will. And, as Laplace proudly announced to Napoleon, there was no room for God either.

Of course, even for an eighteenth-century person blissfully ignorant of quantum mechanics, there were several problems with this argument. Laplace probably meant his assertion more as an allegory than as a serious metaphysical statement.* However, it certainly captures the spirit of the times.

As we make the transition into the nineteenth century, rapid technological innovation emerges as the glorious heir to this scientific tradition, fueling the Industrial Revolution and bringing the belief in the mechanistic approach to nature to an almost euphoric state. By then physics was broadening to include the study of heat and electricity, which, together with Newtonian mechanics, eventually formed the so-called classical worldview.

Carried away by their success, several physicists, most notably Scotland's Lord Kelvin, boasted by 1900 that most of the work in physics had been completed, only minor details being left for future generations. However, for theories, as for people, the danger of

---

*For example, it would be impossible for an intelligence to locate all entities at once, as it takes time to measure their positions—unless this intelligence were omnipresent and omnipotent, concepts Laplace wouldn't have liked very much.

success is overexposure. As technology progressed, so did the quality of laboratory experiments, allowing scientists to probe deeper and deeper into the nature of physical phenomena. Unpleasant surprises started to crop up, becoming more numerous as the century advanced, which combined to dramatically expose the limitations of classical physics. By the time of Lord Kelvin's death in 1907, contrary to his expectations, a radically new conceptual foundation for physics was painfully emerging that would eventually lead to the birth of a brand-new cosmology.

THE NEBULAR HYPOTHESIS

Having mastered the principles of how the Universe operates within Newtonian physics, scientists found themselves increasingly preoccupied with how those principles arose. What could be the role of a Creator in a mechanistic universe? For Newton, God's constant presence insured the stability of an infinite universe. This was the point of view of the so-called *Theists*, who attributed to God the dual job of creator and ever busy mechanic, fixing things here and there as needed. The German mathematician Leibniz would mockingly say that Newton's God was very inefficient, since He had to constantly interfere with His creation. A more efficient God would have created a self-sufficient universe, capable of regulating itself through its own internal mechanisms.

With the rise of Newtonian mechanics, a God who was constantly interfering with the Universe became less and less necessary. One of Newton's arguments for a Designer had been the mysterious fact that all planets orbit the sun in the same direction, while also lying approximately in the same plane. Why should this clear display of order exist if not as the product of some intelligent Creator? A hundred years later, though, Laplace formulated an evolutionary model for the formation of the solar system, which explained some of the features that, in Newton's time, were thought to be compelling arguments for a Designer. Another gap was seemingly closed, forcing the Newtonian gap-filling God to retreat farther into the background.

Laplace built upon previous arguments advanced by the German philosopher Immanuel Kant, who in 1755 theorized that a rotating gas cloud would flatten into a disk as it contracted under its own gravity. Kant was fascinated not only by the Milky Way, but also by other smaller, fuzzy objects that glowed faintly in the night sky. These objects were collectively known as "nebulae," from the Latin for "cloudy." A particularly famous nebula can be seen with the naked eye in the constellation Andromeda. In fact, it is the farthest object visible to us without a telescope, at a distance of about two million light-years.* In the southern tropics where I grew up, the Small and Large Magellanic Clouds are a breathtaking sight. They are considered our satellite galaxies, being located a mere two hundred thousand light-years away.

Galileo had believed that all nebulae were just groups of stars that appear fuzzy because of their enormous distance from us. Kant agreed with Galileo, but elaborated further. Just as stars are grouped by their gravitational attraction to form these distant nebulae, so do the nebulae group together. They form even larger clusters, which he called "island universes." Kant believed in a hierarchical universe, created "in accordance with the infinity of the great Builder."

Like Kant, Laplace conjectured that the solar system formed as a huge, whirling cloud of gas condensed under the attraction of its own gravity. As the cloud gradually flattened into a disk, concentric rings of material broke off due to rotational forces, while most of the mass fell into the center, eventually forming the Sun. The rings then condensed into the individual planets. According to Laplace, at a certain point in its evolution, the solar system looked very much like a giant Saturn. This dynamical mechanism for the formation of stars and their planetary systems became known as the "nebular hypothesis."

The nebular hypothesis was a devastating blow to the Theists who followed in Newton's footsteps. Why should we invoke God to create the observed order in the solar system when simple mechanical arguments would do? Pounded by arguments like this, the God of the Theists slowly gave way to the watchmaker God of the *Deists* who, after creating the Universe, left it to unwind under the strict control of the laws of physics.

---

*A light-year is the distance traveled by light in one year, roughly six thousand billion miles. Thus, the light we receive now from Andromeda left it about two million years ago! The deeper we look into space, the farther back we look in time.

The Deists worked out a compromise between maintaining a belief in God and the rational principles emerging from the Enlightenment. God becomes the first cause and the creator of immutable and universal laws, which we discover through a scientific approach to nature. Since God does not actively intervene in the world, Deists did not accept the existence of miracles. All that is supernatural left in the Universe is the mystery of its creation and the divine conception of the universal laws that rule its behavior. The laws of physics are the work of God, and the job of the scientist is to unveil these laws.

William Paley, an English theologian whose books on Christianity and science were widely read during the nineteenth century, provided several arguments in support of Deism. With some liberties, this is how one of his most famous arguments goes. Suppose, wrote Paley, that while "crossing a heath" one day, you stumble upon a watch lying on the ground. Even if you had never seen a watch before, after a quick examination you would realize that the watch was the product of clever craftsmanship. You would infer that someone had built this object for a purpose, even though this purpose might not be immediately obvious to you. After tinkering with the watch for a while, and assuming it was still working, you would marvel at the discovery that indeed the object had a purpose, being a device built to keep time reliably.

Now look around yourself, Paley would argue, and admire the endlessly detailed sophistication of nature. How can you possibly believe that all this working complexity, so mysteriously efficient, is not the work of a Designer? How could there be no purpose in the awesome sophistication of the Universe? The same excitement you felt when you realized the purpose of the watch, a scientist feels whenever she is lucky enough to unfold one more layer of the cosmic mystery. Nature is the creation of the Clever Watchmaker, while science is the blueprint of the clockwork mechanism.

The problem with this argument, as noted by David Hume and, more recently, by the physicist and writer Paul Davies, is that it works by analogy; the watch had a maker, and so the Universe should have one too. While Paley's argument skillfully lends support to the idea of a Designer, it is certainly not proof. Even if we are tempted to consider it compelling, what may appear today as evidence for cosmic design may be explained away tomorrow by more sophisticated scientific arguments. "So," you may ask impatiently, "will we ever be able to answer this fundamental question?" What's a scientist to do? I am afraid I don't know how to answer that, but if it's any consolation, I don't believe anyone else can either. Although we cannot rule out the possibility that a definitive proof for the existence of a Cosmic Designer is already imprinted somewhere in nature, patiently waiting to be discovered by us, we also cannot rule out the possibility that such proof will never be accessible to us through science. Or that it even exists. Unless we *believe* in it. There are many answers to this question, scientific or not, each partially satisfying our need to understand the ultimate origin of all things.

At this point, all we can do is follow Paley and speculate, based on our own personal beliefs. It is not obvious to me that the beauty we often find in nature cannot be the result of purposeless accidents. But then, it is also not obvious to me that it should. What confuses the issue is that oftentimes the beauty of nature is the result of a compromise between randomness and optimization. Consider, for example, snow crystals. Their beautiful hexagonal (sixfold) symmetry prompted Kepler to write a remarkably prescient essay in 1611 pondering on the possible reasons behind such arrangement.* We now know that the hexagonal symmetry of snow crystals stems from the arrangement of the oxygen atoms in the water molecules forming the ice crystals. In this case, *the emergence of beauty is controlled by physical laws.* And yet, no two snow crystals are identical. Their seemingly endless diversity is caused as freezing water droplets search for the most efficient ways to dissipate heat into their surroundings, being sensitive to local, unpredictable variations in temperature and humidity. *The diversity of beauty is caused by randomness.*

Inspired by the beauty of snow crystals and their complicated formation process, we can focus our discussion on the nature of physical laws. Are *they* evidence for Design? It is very tempting to say

---

*Kepler's approach was far ahead of his time. He not only asked the question "why six?" for the first time (just like the number of known planets then), but also examined and discarded several arguments that attempted to answer this question, finally leaving it as a challenge to future natural philosophers. He anticipated the importance of understanding pattern formation, while recognizing it as a problem for future scientists.

that physical laws are "intelligent." After all, it is because we are intelligent that we can uncover the veiled mechanisms by which nature operates, expressing them in terms of laws. But this position can be dangerous. The fact that it takes intelligence to uncover physical laws does not necessarily imply that they are a product of intelligent Design. Unless, of course, we *believe* that our intelligence itself is not a consequence of random natural selection, but the product of Design.

Is it a fundamental property of being intelligent that we must find intelligence all around us? After all, if pattern recognition is one of our brain's foremost powers, shouldn't we be expected to find intelligence in a natural world full of complex patterns? Are we trapped within our own modes of functioning to the point of blindness? Or is our functioning a premeditated outcome of Design? Until we properly understand the origin of our own intelligence, it may be premature to ascribe intelligence to the Universe.

Let us leave metaphysics aside for a while and return to our discussion of nebulae. Their discovery and enigmatic properties marked a turning point in the history of astronomy. If, as Galileo and Kant hypothesized, they were just agglomerations of many stars, more powerful telescopes should be able to resolve their fuzzy glow into their pointlike components. And so the race for the biggest telescope began. As telescopes became more powerful, the list of observed nebulae started to grow. Still, the mystery of their true nature persisted for a long time. In fact, only in the early 1920s did it become clear that the Andromeda "nebula" was another galaxy and not part of our own Milky Way! Also around the same time astronomers accepted the fact that the solar system was not at the center of the Milky Way, yielding the last victory in the Copernican revolution. As my colleague and friend Rocky Kolb has remarked, "The study of nebulae attracted some of the greatest astronomers from the time of Galileo until today, and confounded most of them."

One of the reasons for the confusion is that the objects that were generically called nebulae in the past comprise objects of very distinct natures. There are:

1. Diffuse nebulae, huge gas clouds that are illuminated by the light of a nearby star or stars

2. Planetary nebulae, which are thin shells of gas blown off by a single star
3. Star clusters, which may be called "open clusters" if they have relatively few stars, or "globular clusters" if they have millions of stars
4. Finally, galaxies, which can have anywhere from hundreds of millions to ten thousand billion stars.

The first systematic catalogue of nebulae was compiled by Charles Messier in the 1780s. Out of 103 nebulae (42 discovered by Messier himself), 7 are diffuse nebulae, 4 are planetary nebulae, 28 are open clusters, 29 are globular clusters, 34 are galaxies, and one is a binary star system, two stars rotating around each other. The skies had become very busy, populated by a variety of objects completely unknown to earlier sky gazers.

Perhaps no other eighteenth-century astronomer knew the heavens with the precision of William Herschel, who was fond of comparing the heavens to a garden "which contains the greatest variety of productions." With the obsessive zeal of a half-crazed botanist, Herschel set off to map the skies, armed with one of the largest telescopes of his times. In 1788, he had a telescope with a mirror four feet in diameter. By the time of his death in 1822, he had discovered a new planet, Uranus; had produced a catalogue with 2,500 nebulae; had pioneered the study of stellar astronomy; and had attempted, after many false starts, a first classification of the different kinds of nebulae, coining the terms *planetary nebula* and *globular cluster*. He had glimpsed, far more than any before him, the enormous diversity of the heavens.

Raw telescopic power revealed a wholly new dimension to the night sky—its depth. By the mid-nineteenth century, most astronomers were (wrongly) convinced that *all* nebulae were just groups of stars. If nebulae were indeed groups of stars held together by their gravitational attraction, just how far away were they? How much telescopic power was necessary to solve this enigma once and for all?

As in many instances in the history of science, raw power is not always the answer. Sometimes a fresh new idea is needed, one that will override long-held beliefs, settling questions that would have remained open otherwise. In the case of the nebulae, this new idea was to use a new instrument in the study of astronomical objects, the spectroscope.

A *spectroscope* is a device that separates the light coming from a source into its components, very much as Newton's prism separated the sunlight into the seven colors of the rainbow. The added ingredient of a spectroscope is a narrow vertical slit that is placed between the source of light and the prism. (Instead of a prism, a *grating* can be used, a grooved transparent surface.) What is revealed is known as the *spectrum* of the source of light.

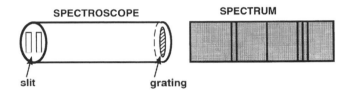

A spectroscope and a spectrum: Light from a source goes through the slit, hitting the grating. The resulting spectrum has a few black lines.

In the early 1800s, Joseph Fraunhofer, a young German optician, came up with a brilliant idea. Why not point a spectroscope at the Sun? Fraunhofer had been using the spectroscope in order to select monochromatic (single-color) lines to test the lenses he produced. When he looked at the spectrum produced by the Sun, he could hardly believe his eyes. He saw "an almost countless number of strong and weak vertical lines which are darker than the rest of the color-image. Some appeared to be perfectly black." Superimposed onto the rainbow colors found by Newton were several dark lines at precise locations. As he scanned from the deepest blue to the strongest red, specific lines of color, that is, where the black lines appeared, were missing!

Fraunhofer catalogued hundreds of those black lines, showing also that the spectra emitted by the Moon and planets were identical to that of the Sun and thereby proving they did indeed just reflect the Sun's light, as Galileo had inferred from his observations of the phases of Venus two hundred years before. But Fraunhofer's discovery also raised some serious questions. What caused the dark lines to appear in the solar spectrum? Where did the missing colors go? And why were only certain colors missing? The final resolution to the mystery of missing lines would have to wait another hundred years, until the enigmatic nature of atomic physics began to be understood. Still, their discovery opened a brand new window to the night sky, which was to profoundly influence the development of astronomy and cosmology.

Although others quickly confirmed the existence of the missing lines in the solar spectrum, only in the late 1850s were important advances made due partly to the development of chemistry and the classification of chemical elements. Between 1855 and 1863, Gustav Kirchhoff and Robert Bunsen heated a series of chemical elements and took their spectra. They soon realized that different elements emit light of different colors, or, more accurately, that each element has its own spectrum. Thus, we can think of a spectrum as a "fingerprint" for chemical elements. If we analyze a mixture of different chemical elements through a spectroscope, the spectrum will reveal the different elements belonging to the mixture.

One night Bunsen and Kirchhoff were working in their laboratory in Heidelberg when they saw a fire raging in the nearby port city of Mannheim. As they pointed their spectroscope toward the fire, they identified the lines of the chemical elements barium and strontium in the flames. Inspired by their discovery, they wondered if they could identify chemical elements in the solar spectrum.

The French physicist Jean-Bernard Foucault had already shown that when a bright light passes through a cool cloud of vaporized sodium, two black lines appear in the resulting spectrum. Furthermore, the two lines correspond precisely to the two bright yellow lines that characterize sodium's spectrum. The sodium cloud selectively "absorbed" its two yellow lines from the bright light.

Kirchhoff then proved that a given element will always emit and absorb the same colors. The rest was "easy"; check which colors were missing in the solar spectrum and compare them to the known spectra of chemical elements.

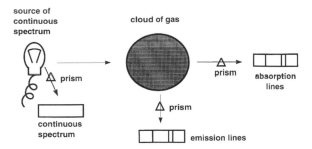

Emission and absorption spectra: A hot cloud of gas produces an emission spectrum with a few black lines (middle).
When the cloud is between a source of continuous spectrum, i.e. without any lines (left) and a spectroscope, it will
absorb selectively, producing an absorption spectrum with
the telltale black lines (right).

By 1861 Kirchhoff had identified a host of chemical elements in
the Sun, such as intense sodium lines, calcium, magnesium, and
iron. These elements, being present in the outer, cooler regions of
the Sun, selectively absorbed their respective spectral colors, generating the missing lines in the solar spectrum. This discovery that
the Sun was made up of chemicals familiar to us on Earth was of
tremendous importance. There was no "fifth essence," as the
Greeks had called the substance that made the heavenly bodies,
just chemical elements that we find all around us. Was this true for
all objects in the heavens, including other stars and the mysterious
nebulae?

William Huggins, a wealthy amateur astronomer, was very excited by the news of Bunsen and Kirchhoff's discoveries. Fitting a
spectroscope to his telescope on Upper Tulse Hill in London, he
patiently pointed it at the bright stars Aldebaran and Betelgeuse. After struggling to disentangle the web of overlapping lines, he correctly identified iron, sodium, calcium, and other elements in the
spectra he collected from these stars. In so doing, he discovered that
other stars were also made of chemicals found in our own solar system, although their individual spectra can vary substantially.

Huggins then decided to look at nebulae. Could he shed some
light on the ongoing debate about their nature? If nebulae were just
agglomerates of stars, he should be able to identify spectra with the

usual proliferation of lines. An 1864 entry to his journal tells the
story:

> [It was with] excited suspense, mingled with a degree of awe, with
> which, after a few moments of hesitation, I put my eye to the
> spectroscope. Was I not about to look into a secret place of cre-
> ation? . . . I looked into the spectroscope. No spectrum such as I ex-
> pected! A single bright line only! . . . The riddle of the nebulae was
> solved. The answer, which had come to us in the light itself, read:
> Not an aggregation of stars, but a luminous gas.

Huggins happened to have pointed his telescope toward a true
gas cloud, and was thus led to the erroneous conclusion that all nebulae were gas clouds. He was wrong in his generalization, but not
about the spectra of gas clouds. For another fifty years the true nature of nebulae would remain elusive.

The discovery of stellar spectra and their relation to terrestrial
chemical elements paved the way to the development of astro-
physics, the branch of physics devoted to the study of cosmic objects. Surely, with better spectroscopes and telescopes, more and
more spectra could be accurately read and interpreted, showing the
many similarities and differences between the various luminous
sources in the night sky. But many fundamental questions remained
unanswered. Why do hot objects emit light? Why do different
chemical elements have different spectra? More generally, what is
light, what is heat, and is there a relationship between the two? The
struggle to answer these questions led to a large part of the work
done in fundamental physics during the nineteenth century. Together, they exposed the power and limitations of the classical
worldview, pointing the way to the new physics of the twentieth century. In order to appreciate why, we must look at the physics of heat,
electricity, and light.

THE ELUSIVE NATURE OF HEAT

There is not a child in the world who isn't fascinated by fire.
When I was growing up in Rio, my family took every opportunity to
escape from the big city to my grandparents' country house, about

seventy miles away. I remember my excitement when, from a proud distance at the head of the dining table, my grandfather announced that it was cold enough to light the fireplace. Granted, for a Brazilian child part of the excitement was that this happened only a few nights a year. But still, as soon as the flames started to consume the wood, nothing could move me from my position four feet from the action. My cousins and I would poke the wood, causing big burning ashes to fly up the chimney, much to my grandmother's despair. "The carpet! Watch out for the carpet! Your hands! Please be careful . . . You will wet your beds tonight, you crazy kids!" (I never quite understood the connection between fire and wet beds, which is quite widespread in Brazil. Anyway, I am glad to say that my grandmother's cursing never worked on me, although I can't say the same about my cousins.)

Fire has a dual nature, being both dangerous and useful, beautiful and destructive. The fact that burning materials produce heat is responsible, to a large extent, for our survival as a species. Nevertheless, a true understanding of the process of combustion and the nature of heat was to elude scientists until well into the nineteenth century. The first attempt at understanding how things burn was put forward by the German chemist George Ernst Stahl (1660–1734), who postulated that burning resulted from the release of a hypothetical element called *phlogiston*. Every combustible substance was made of a combination of phlogiston and the residue left over after the burning.

The great French chemist Antoine Laurent Lavoisier was the first to properly understand the process of combustion as a chemical combination of a fuel with oxygen. Without oxygen, materials don't burn. Lavoisier proved this fact in a series of brilliant experiments that revolutionized chemistry. In one of them, he borrowed a few diamonds from a Parisian jeweler, placed them in an airtight clay vessel, and proceeded to bake them. Much to the relief of the jeweler, Lavoisier showed that, in the absence of air, the diamonds didn't burn. He also showed that, in the process of burning and of any chemical transformation, the total mass of the reacting substances is conserved. There was no need to invoke phlogiston as a hypothetical element in order to understand burning. In 1789, he proposed the law of mass conservation:

We must lay down as an incontestable axiom, that in all the operations of art and nature, nothing is created; an equal quantity of
matter exists both before and after the experiment. Upon this principle, the whole art of performing chemical experiments depends.

However, the nature of heat remained obscure. We all know
that heat flows from hot to cold objects; this is why we can tell when
someone has a fever by placing our hand on their forehead, or why
a hot pot of soup cools when we remove it from the flame. The simplest and most intuitive explanation is that heat is a sort of invisible
fluid, which flows spontaneously from hot objects to colder ones.
This, in fact, was the assumption of the *caloric hypothesis*, which was
advocated by none other than Lavoisier himself. In order to keep
things consistent with his law of mass conservation, the caloric fluid
was supposed to be massless, and its total amount in the Universe
conserved. The only way to detect the presence of caloric was when
a difference in temperature between two or more bodies induced
the flow of heat. It was an ingenious idea, which, although wrong,
allowed for a tremendous amount of progress in the study of heat
and its possible technological applications.

Of all technological innovations that appeared in the later eighteenth century, none is more closely associated with the Industrial
Revolution than the steam engine. The mechanization of production, both in the agricultural and manufacturing sectors of the
British economy, became synonymous with progress. When the
Scotsman James Watt patented the first really efficient steam engine
in 1769, the race was on to improve the efficiency of steam-powered
machines—more mechanical work generated with the same amount
of coal.

As steam propelled the Industrial Revolution forward, efficiency
was equated with the economic ideal of surplus. Better-built engines
meant more output for less fuel, and thus more money in the bank
accounts of the emerging class of wealthy industrialists. The question was: How efficient could a steam engine be made? The efforts
to answer this question gave rise to a new branch of physics, the
study of heat, or *thermodynamics*.

Using the caloric hypothesis, the French engineer Nicolas-
Léonard-Sadi Carnot (1796–1832) clarified some of the physical

principles behind the steam engine. He noted that a steam engine shared several features with a waterwheel. As water falls on the paddles of the waterwheel, causing it to turn, it is possible to move different kinds of machinery. Before we proceed with Carnot's waterwheel analogy, we should stop to discuss how physicists understand energy.

In mechanics, energy is divided into two possible types, potential and kinetic. *Kinetic energy* is the energy carried by objects in motion, while *potential energy* is energy that is somehow stored in them. The wonderful thing is that one can easily be turned into the other. A simple device for studying how potential energy can be converted into kinetic energy is the slingshot. After placing a rock on the rubber band, we pull it backward, causing potential energy to be stored in the elastic band. Releasing the band propels the rock forward, changing the potential energy into the kinetic energy of the rock's motion. A gun works on the same principle: The chemical potential energy stored in the gunpowder is converted into the bullet's kinetic energy.

Here is another example: A diver on a platform possesses stored gravitational potential energy. The higher the platform, the more potential energy she stores by climbing on it. In fact, anything that can fall has stored gravitational potential energy. And the higher the climb, the harder the fall. That is, the more potential energy you store going up, the more kinetic energy you will have going down.*

Let's now go back to Carnot's waterwheel analogy for a steam engine. As the falling water strikes the paddles, its gravitational potential energy is converted into kinetic energy. The higher the water source, the more kinetic energy it has when it hits the paddles. In the end, gravitational potential energy is converted into the mechanical energy of the turning wheel. Carnot reasoned that a steam engine works in a very similar way. Just as falling water moves a waterwheel, flowing heat will move the steam engine. In order to improve

---

*This is only strictly correct in the absence of air. A body falling in the presence of air will reach a "terminal velocity" which is constant. I would not, however, recommend that you try this experiment on your next airplane trip; for a human body falling vertically, the terminal velocity is about 200 miles per hour.

the efficiency of the steam engine, we must increase the temperature difference between the source of heat and wherever it is flowing to, just as we must raise the water higher to increase the mechanical output of the waterwheel.

Carnot also realized that, although compelling, the analogy was not perfect. In a steam engine, the temperature difference is between the steam and the surroundings. How could the temperature difference be increased if steam is at the boiling point of water, 100 Celsius degrees (212 Fahrenheit degrees)? Carnot pointed out that in order to increase the temperature difference, steam had to be produced at higher pressures, just as in a pressure cooker; keeping its volume constant, a gas at higher pressures is hotter.

Carnot published his ideas in 1824, in a book called *Réflexions sur la puissance motrice du feu et sur les machines propres à développer cette puissance*, which can be translated as "Thoughts on the motive power of fire, and the adequate machines for developing this power." However, only with the work of William Thomson (later Lord Kelvin) and of the German Rudolf Clausius (1822–1888), was the significance of Carnot's work finally understood twenty years later. Inspired by Carnot's arguments, they realized that in a real engine some of the heat was used to boil the water, some was always lost to the surroundings due to friction, and some was just lost; there was no such thing as a perfect engine. As the engine repeated its cyclic motion, transforming water into steam, which then moved a mechanical gadget before condensing back into water, it was not possible to recover all the heat released by the cooling water to restart the process. Some extra fuel was needed to keep the engine going, compensating for an unavoidable heat loss. This led them to conclude that while it is fairly easy to convert mechanical work into heat (this is what you do when you rub your hands to keep them warm), the reverse is much harder. (Imagine what life would be like if heat made you rub your hands.) Only a fraction of the heat in a system is "useful heat," that can be converted into organized mechanical work.

A simple "thought experiment" illustrates what I mean by useful heat. Consider a transparent cylinder, with a piston on top of it that can move up or down (without friction) as in the figure opposite. A thermometer is attached to the cylinder so that we can measure the temperature of the air in its interior. Assume that no heat is allowed to flow out of the cylinder. (Aren't thought experiments wonderful?) Now let's heat up the cylinder with an oil lamp. As the flame burns by consuming the oil, the air in the cylinder gets hotter and expands, pushing up the piston. Underlying this simple operation is the *first law of thermodynamics*: The total amount of energy in an isolated system (the cylinder, its contents, the oil lamp, and the air around it) must be constant; the total energy must be the same, before and after. The chemical energy originally stored in the oil is equal to the energy used to heat up the air around it and in the cylinder, plus the gravitational potential energy of the elevated piston.

Now cool the cylinder so that the piston gets back to its original position. Put a pendulum inside the cylinder and make it swing. As the pendulum swings, it heats up the air by friction, causing the piston to move up just as if the cylinder were being heated up by a flame. (Remember, this is a thought experiment.) When all the pendulum's mechanical energy is changed into heat, it will stop in a vertical position.

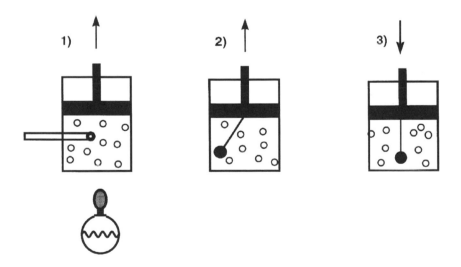

As the oil lamp heats up the air inside the cylinder, the piston goes up. The piston also goes up as the mechanical energy of an oscillating pendulum heats up the air by friction. Pushing the piston down heats up the air, but does not make the pendulum oscillate again. The circles are (greatly enlarged!) air molecules.

All the mechanical energy of the pendulum was used in heating the air in the cylinder and in elevating the piston. This is again an expression of the first law of thermodynamics, since mechanical energy is turned into heat; you can increase the temperature of a gas either by heating it or by "agitating" it through mechanical means. During the 1840s, the British physicist James Joule measured, in a sequence of landmark experiments, the mechanical equivalent of heat, that is, how much heat is generated by a given amount of mechanical work. Heat is just another form of energy.

Now comes the crucial part: Push the piston down to its original position. As the pressure is increased inside the cylinder, the air temperature goes up. In a perfect world, you would expect that the energy stored into heat would cause the pendulum to start moving again. But of course it doesn't. Once organized mechanical motion is dissipated into disorganized heat, you can't get it back.

As Clausius was pondering how to quantify the usefulness of heat, he came across the idea of *entropy*. Entropy can be understood as a measure of a system's ability to do organized work. A system with low entropy has a higher ability to do organized work than a system with higher entropy. A typical signature of an irreversible process is the increase of entropy. The experiment above is an example of an irreversible process. The system spontaneously generates heat from mechanical energy (the slowing down of the pendulum by friction), but it does not spontaneously generate mechanical motion from heat (heating up the air doesn't make the pendulum oscillate again).

Heat is disorganized energy, and it is hard to tame it into generating organized behavior. As a consequence, in the evolution of any system, the final state will be less ordered (have higher entropy) than the initial state. This is what is known as the *second law of thermodynamics*. Here are a few familiar examples of irreversibility in action. A sugar cube melts in a hot cup of coffee, but does not reassemble itself. An omelette does not revert into an egg. Perfume flowing out of an open bottle does not flow back into it. Warm water does not separate into cold and hot.

In other words, the second law states that in any *isolated* physical system, entropy always increases. Isolated here refers to a system that is not allowed to freely absorb energy from the environment. In an open (as opposed to isolated) system, entropy is allowed to decrease. This is why complex organized structures can emerge locally, such as sugar cubes, clean homes, organic macromolecules, and ultimately life itself. In order to develop, life forms feed on the environment, leaving behind waste products. Although order emerges locally, entropy increases globally. Disorder always wins in the end. Sounds depressing? Well, think of the opposite alternative; a world of constant entropy is a world with no change; everything would either be static or perfectly cyclic, always coming back precisely to the point it started, in endless perpetual motion. This, I think, is a much more depressing alternative.

Irreversibility is intimately related to the direction of time. If I make a movie of a sugar cube melting in a cup of coffee and play it backward, you would know such a thing is impossible. The melting of the sugar cube implies a direction of time that is irreversible. On the other hand, if I make a movie of a soap bubble floating around and played it backward, you wouldn't know the difference; the bubble's motion is reversible (unless it popped). Why is there irreversibility for the melting sugar cube, but not for the bubble's motion? The answer is in the complexity of the system under question. For the purpose of this discussion, it is not important to define what complexity really means. In fact, there are many possible definitions of complexity, none being entirely satisfactory. In principle, the sugar crystals could reassemble spontaneously into a cube, but the probability for this is so astronomically small as to be completely negligible; it simply won't happen. Irreversible motion is a consequence of the complexity of natural systems. The more complicated a system is, the more improbable it is that the system can revert to its original state.

The concepts of entropy and irreversibility introduce several new ingredients into physics: probability and microscopic behavior. Thermodynamics deals exclusively with observable properties of systems, such as their pressure, volume, and temperature. It doesn't really explain *why*, for example, heating a gas increases its temperature. By the mid-nineteenth century, the only available explanation was still based on the caloric hypothesis, that caloric (heat) cannot be created or destroyed, but just passed from a hotter object to a colder one. But it was becoming increasingly obvious that this explanation just wasn't enough. If that were true, where does the heat come from when we rub our hands together? The two hands are at the same temperature, and yet rubbing them together creates heat.

Proponents of the caloric hypothesis would claim that the action of rubbing caused caloric fluid to "bleed" from the material, thus liberating the observed heat. If that was so, one could imagine that at some point, the amount of caloric heat in a given material would be exhausted, and no more heat could be produced by friction. This theory clearly could not be upheld, and Benjamin Thompson (1753–1814), an American expatriate later to be known as Count Rumford, became its professed opponent. After serving as a loyalist officer in King George III's army fighting in the American Revolution, Rumford left England for Munich, where he was promoted to general by the elector of Bavaria. While in Munich, he oversaw the boring of cannons, a perfect laboratory situation to study the generation of heat by friction. Using water to cool the drill and the cannon, he marveled at the tremendous amount of heat being liberated, which would quickly bring the water to a boil and keep it boiling for as long as the boring continued. He pointed out that the amount of heat being generated by friction "appeared evidently to be inexhaustible," writing, in 1798:

> Anything which any insulated body, or system of bodies, can continue to furnish without limitation, cannot possibly be a material substance; and it appears to me extremely difficult, if not quite impossible, to form any distinct idea of any thing, capable of being excited and communicated in the manner the Heat was excited and communicated in these experiments, except it be Motion.

Making one of the first suggestions that heat is related to motion, Rumford was convinced that the caloric hypothesis was wrong. After Lavoisier was tragically beheaded during the Reign of Terror, Rumford wrote to his widow, whom he was about to marry, "I think I shall live to drive caloric off the stage as the late Monsieur Lavoisier drove away phlogiston. What a singular destiny for the wife of two Philosophers!"

Even before Rumford's experiments, there were other proponents of the idea that heat was related to motion. In 1738, Daniel Bernoulli (1700–1782) proposed a microscopic model for gases that is remarkably close to the so-called *kinetic theory* developed during the late nineteenth century, which would finally elucidate the true nature of heat. Assuming that a gas consists of innumerable molecules in rapid random motion, Bernoulli was able to show that when enclosed in a container, the pressure of the gas comes from collisions of these molecules against the walls of the container. If we halve the volume in a container by pushing down on a piston while keeping its temperature constant, the pressure doubles. Bernoulli proposed that the increase in pressure was due to squeezing the gas molecules into a smaller volume; as their density increases, so does the number of collisions with the piston. It was a remarkable result that, nevertheless, was to be ignored for over one hundred years.

The next great contribution to a microscopic theory of heat came in 1845, when the British physicist John James Waterston submitted a paper to the Royal Society in which he outlined the relations between the temperature and pressure of a gas and the average velocity of its molecules. He obtained two crucial results:

1. The temperature of a gas is proportional to the square of the average velocity of its molecules, and
2. Its pressure is proportional to the density of the molecules times this average speed.

Thus, macroscopic properties of gases, such as their temperature and pressure, could be understood in terms of the motions of their microscopic constituents. As a gas is heated, the increase in temperature is due to an increase in the average velocity of its molecules. Heat and motion are indeed intimately related!

Unfortunately, Waterston's paper was rejected by two referees and was lost in the archives of the Royal Society. One of the referees wrote in his report that "the paper is nonsense, unfit even for reading before the Society," while the other stated that the paper "exhibits much skill and many remarkable accordances with the general facts . . . but the original principle is . . . by no means a satisfactory basis for a mathematical theory." Behind this criticism was a strong prejudice against the atomic, or corpuscular, theory of matter that was to endure until the turn of the century. Most physicists couldn't accept the existence of postulated objects that could not be seen, even if they successfully explained so many of the observed properties of gases.

Prejudice apart, the corpuscular theory of gases gained a tremendous boost with the publication in 1860 of a brilliant paper by James Clerk Maxwell, entitled "Illustration of the Dynamical Theory of Gases: On the Motion and Collision of Perfectly Elastic Spheres." Assuming that the molecules were hard spheres moving under Newton's laws and colliding with each other without losing kinetic energy (hence the term *elastic*), Maxwell obtained several results reinforcing the ideas of Bernoulli and Waterston.

Maxwell's results were further generalized by the great Austrian physicist Ludwig Boltzmann (1844–1906) in a monumental work on kinetic theory. In it he showed that statistical arguments based on the motions of gas molecules could be used to derive the laws of thermodynamics. In a crucial departure from the Newtonian worldview, Boltzmann showed that it was hopeless, indeed unnecessary, to follow the motions of individual molecules in order to explain the macroscopic properties of gases. It was their cumulative motion that was responsible for the macroscopic properties of gases measured in the laboratory. And their cumulative motion was accurately described by statistical arguments.

Despite its explanatory power, the atomistic foundations of the kinetic theory were still seen by many as merely a conceptual device. In 1883, the famous physicist and philosopher Ernst Mach wrote:

> Atoms cannot be perceived by the senses; like all substances, they are things of thought. Furthermore, atoms are invested with properties that absolutely contradict the attributes hitherto observed in bodies. However well fitted atomic theories may be to reproduce certain groups of facts, the physical enquirer who has laid to heart Newton's rules will only admit those theories as provisional helps, and will strive to attain, in a more natural way, a satisfactory substitute.

Toward the end of the century, Boltzmann found himself increasingly isolated in the fight against mounting criticism to the kinetic theory. His feelings were forcefully expressed in the foreword to the second volume of his book, written in 1898:

In my opinion it would be a great tragedy for science if the theory of gases were temporarily thrown into oblivion because of a momentary hostile attitude toward it. . . . I am conscious of being only an individual struggling weakly against the stream of time. But it still remains in my power to contribute in such a way that, when the theory of gases is again revived, not too much will have to be rediscovered.

Suffering from severe depression and in poor health, Boltzmann committed suicide in 1906, just two years before the experiments of the French physicist Jean Perrin confirmed many of his ideas. Although we will never know how much of his despair was caused by the negative reaction to his lifelong work, his death remains one of the most painful episodes in the history of science. His ideas, however, have been more than vindicated. The kinetic theory unveiled, once and for all, the true nature of heat. All observable properties of gases can indeed be understood in terms of the motions of molecules, individually dancing to Newton's laws but collectively described by statistics. From pre-Socratic philosophy to testable theory of matter, atomism makes a triumphal reentrance into the physics arena.

WAVES OF LIGHT

Storms conjure up all sorts of ancient fears. You may be a well-informed, late-twentieth-century person, in touch with the world through CNN and the Weather Channel, fully aware of nature's unruly manifestations. But consider this situation: One beautiful summer afternoon, as you are leaving your office to go home, you notice a pleasant breeze blowing from the east. Soon the pleasant breeze turns into a furious wind, bringing in clouds from, it seems, all directions at once. An hour later, you don't know if it's dark because of nightfall or because of the massive clouds. You remember, with a shudder, that at this time of the year it stays light until nine or so. You look up and wonder how heavy can clouds possibly get before they discharge. There is a brief moment of complete stillness. And then it starts.

Your house is under the attack of a heavy electrical storm.

Lightning is flashing all around you, painting your walls with elusive ghostly blue shadows. Continuous thundering, loud to the point of deafening, eerily shakes your bed (yes, somehow you find yourself under the covers by now) and your wits. Water is pouring down from the heavens, without an indication of stopping. The power goes off (of course), you hear a terrible explosion, followed by a loud cracking sound; your beautiful two-hundred-year-old pine tree falls, consumed instantly by flames. All you can think about is the lightning rod while imploring to yourself, "Please, *please* work!"

At least you *have* a lightning rod, or some other way to diffuse the potential damage of lightning. Imagine the havoc caused by electric storms before the invention of the lightning rod. We can thank none other than Benjamin Franklin (1706–1790) for this clever invention. In the summer of 1752, during a storm like the one I described above, Franklin decided to prove his thesis that lightning was related to electricity. As lightning started to appear, Franklin and his son boldly went outside to fly a silk kite. They had attached a metal key to the string, and noted that when "electrical fire" hit the wet kite, sparks flew off the dangling key. But if the kite was connected to the ground, the lightning would harmlessly discharge itself. The lightning rod was born.*

During the mid-eighteenth century, electricity was thought, just like heat, to be a fluid. Actually, because it was known that electrified objects can both attract and repel each other, electricity was thought to be made of two fluids. Apparently, Franklin was not aware of this two-fluid model of electricity and proposed a simpler, and more correct, one-fluid model. The so-called electric fluid was supposed to be present in every material body. When two bodies are rubbed against each other, some of this fluid is displaced. If an object gains fluid, it becomes positively charged; if it loses fluid, it becomes negatively charged. For example, if a glass rod is rubbed with a silk handkerchief, the rod becomes positively charged, while the silk becomes negatively charged. Note that this simple fluid theory implies that charge can neither be created nor destroyed, but just

---

*Franklin was not just a clever inventor, but also a very lucky one. A year later, G. W. Richmann tried the kite trick in St. Petersburg, Russia, and was struck dead by lightning in a matter of minutes.

transferred. Like energy, the total electric charge in a system is conserved, a natural law of fundamental importance.

Franklin's theory also makes sense, within reason, from a modern point of view. We now know that matter is composed of atoms, and that atoms are composed of positively charged nuclei and negatively charged electrons. Left alone, matter is neutral. However, when materials are rubbed against each other, they lose or gain electrons, becoming electrically charged. If electrons are gained, the material picks up an extra negative charge. If electrons are lost, the material picks up an extra positive charge. Thus, the only "flaw" in Franklin's theory was his choice of charge for the electric fluid.

The next big step in the study of electricity was the measurement of the electric force between two electrified bodies. Again, Franklin played an important role, even if by this time he was more interested in politics than science. As the representative of colonial Pennsylvania, Franklin took advantage of his residence in England to participate in the meetings of the Royal Society, to which he had been elected in 1756. In one of these, he mentioned to Joseph Priestley his peculiar discovery that a small cork ball hanging from a string is strongly attracted by a charged metal shell if brought close to it, but not if it is placed inside the shell.

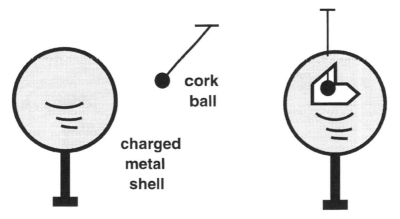

According to one of Franklin's experiments, a cork ball is attracted to a large charged shell. However, when placed inside the shell, the cork ball experiences no force of attraction at all.

Priestley immediately noticed an analogy with the gravitational force: A small mass placed outside a large massive shell is attracted to it, but not if it is placed in the shell's interior. This result, Newton had demonstrated in the *Principia*, was a consequence of the inverse square law of gravity. Roughly, the small mass within the shell is being attracted by the outer shell in all directions, and the sum total of all forces on the small sphere exactly cancels.

Do charges also attract and repel each other by means of an inverse square law? Hearing Priestley's arguments, the English chemist Henry Cavendish set up a delicate experiment to put them to the test. He surrounded an insulated charged sphere with a larger uncharged metallic shell. The two objects were then connected by a wire, which let the charge freely flow between them.

After removing the outer shell, Cavendish discovered that the interior sphere was neutral; all the charge had migrated to the outer shell. Using Newton's calculus, Cavendish demonstrated that this was possible only if the force law between charged bodies varied with the inverse square of their distance, just like gravity. Curiously, Cavendish never published these results, and his epochal discovery was to remain unknown for another hundred years. It was in France that measurements of the inverse square law for electricity were first made public, after the careful experiments of Charles-Augustin de Coulomb (1736–1806) in 1785. To date, the inverse square law for electric charges is known as Coulomb's law.

The fact that electrical and gravitational forces show such a remarkable similarity points to a deep simplicity in the way nature operates. Whenever a physicist is faced with an unexpected result like this, she immediately starts searching for a deeper level of explanation. If the two forces operate under the inverse square law, this must be an obvious consequence of some unknown underlying principle. It is as if some new physics, previously undreamt of, is lurking behind the scenes, insinuating itself through a few meager hints. Inspiring as it may be, the search for such underlying principles can also be very elusive and frustrating. In the case of the relationship between electricity and gravity, the search is still going on, after having defeated some of the greatest minds of all time, Einstein's included. But defeat only increases the challenge and the rewards of a potential discovery, unless, of course, "intuition" turns into blind obsession, and the challenge into a major waste of time. But how is one to know when to stop?

An inverse square law of electrical forces raised an old specter from the grave: the concept of "action at a distance." How could two charged objects influence each other through empty space? The same was true for magnetism, this other mysterious force that had inspired Kepler on his speculations about the causes of planetary motions. By the early nineteenth century magnetism was beginning to look even more like electricity, since magnetized materials also both attracted and repelled each other. A series of discoveries concerning the behavior of electric and magnetic forces were to promote dramatic changes in the established Newtonian worldview. By the late nineteenth century, action at a distance would be replaced by the novel concept of *field*. Electricity and magnetism would be shown to be manifestations of a single electromagnetic field.

Chance helps those who are well prepared. Although it is true that luck has played a role in several scientific discoveries, it is also true that plain luck is not enough. Usually, a "serendipitous" discovery happens when a scientist has spent much time and effort carefully looking for something. Would Fraunhofer have discovered the dark lines in the solar spectrum had he not pointed his spectroscope at the Sun? And so it was with the first link in the chain leading to the discovery of electromagnetism.

During the winter of 1820, Hans Christian Oersted (1777–1851), a Danish physicist, and good friend of another famous Hans Christian more interested in fairy tales than science, was teaching a class on electricity and magnetism. Oersted had long suspected a connection between electricity and magnetism, inspired as he was by Kant's belief in the unity of natural phenomena. In fact, back in 1813 Oersted wrote:

One has always been tempted to compare the magnetic forces with the electrical forces. The great resemblance between electrical and magnetic attractions and repulsions and the similarity of their laws necessarily would bring about this comparison. An attempt should be made to see if electricity has any action on the magnet as such.

Preparing for a lecture, Oersted had all sorts of apparatus on his table, from voltaic cells (batteries)* and wires to magnets and compasses. While he was demonstrating how a voltaic cell could be used to generate an electric current, he noticed, to his amazement, that whenever a current flowed through the wire, the needle of a nearby compass moved! But how could that be? The only thing that can deflect a compass needle from a distance is a magnetic force. Oersted quickly realized that it was the electric current that was generating the magnetic force. Since electric current means electric charges (or, more appropriately for the times, fluid) in motion, Oersted deduced that the motion of electric charges generates a magnetic force. There was indeed a deep relationship between electricity and magnetism!**

Oersted's discovery caused tidal waves in the European scientific establishment. In France, André-Marie Ampère (1775–1836) and others set up experiments exploring the forces between charge-carrying wires. And in England, the relationship between electricity and magnetism caught the attention of a young laboratory assistant destined to become one of the scientific giants of all time.

Michael Faraday was born September 22, 1791, in Surrey, the son of a blacksmith. He grew up in utter poverty, at times so bad that he had to survive an entire week on a loaf of bread. When he was five, the family moved to London, although with no apparent gain in wealth. As he wrote later, "My education was of the most ordinary description, consisting of little more than the rudiments of reading, writing, and arithmetic at a common day school. My hours out of school were passed at home and in the streets."

Faraday was a self-starter, though. When he was thirteen, he became an apprentice bookbinder, surrounding himself with dozens of books that were the beginning of his scientific education. As he wrote later to a friend:

---

*Voltaic cells were a recent invention coming from Italy, so named after Count Alessandro Volta, who first demonstrated how a battery can keep a steady electric current.

**We now know that magnets and other naturally magnetic materials owe their magnetism to the motion of electric charges at the atomic level. Magnetism is electricity in motion.

It was in those books, in the hours after work, that I found the beginning of my philosophy. There were two that especially helped me, the *Encyclopaedia Britannica*, from which I gained my first notions of electricity, and Mrs. Marcet's *Conversations on Chemistry*, which gave me my foundation in that science.

By the time he was nineteen, whatever extra money he could save Faraday used to conduct experiments on electrolytic decomposition (using electric currents to promote chemical decomposition of substances, for example, water into oxygen and hydrogen). During the spring of the same year, the generosity of a customer made it possible for Faraday to attend four evening lectures by the distinguished chemist Sir Humphry Davy at the Royal Institution. These lectures were to change his life. He took meticulous notes, extended them, and used his skills as a bookbinder to produce a handsome volume, which he then boldly sent to Davy with a request for employment at the Royal Institution.

The Royal Institution had been founded (by Count Rumford) with the noble (and unrealistic) ideal of improving the educational level of the working class through exposure to science. (Better wages and schools would certainly have been more efficient.) Public lectures were to be attended by workers, eager to improve their lives bathed in the shining light of knowledge. However, an average of seventy-plus hours per week of work in miserable conditions left the working classes very little eagerness for science or knowledge in general. The lectures were mostly attended by the very middle-class that sponsored them. But Faraday was certainly a member of the underprivileged working class, and he was requesting help from the Institution. Although he was a shoo-in according to the Institution ideals, Davy initially advised Faraday to stick to bookbinding, since a scientific career did not offer any guarantee of job security, money, or opportunity for advancement. Davy's advice to Faraday is echoed every day in universities all over the world.

Nonetheless, in March 1813 a laboratory assistant was dismissed, and Davy called on Faraday. He was given aprons, fuel, candles, two rooms in the garret of the Royal Institution, and free access to the laboratories. Soon after, the two left London for eighteen months, touring several laboratories and universities in France, Italy, and

Switzerland, where Faraday met some of the great scientists of the day. He learned quite a few things about science, but also about himself. He was never again to leave the simple life of the laboratory for the pomp and glitter of high society. Much later in his career, when offered the presidency of the Royal Society, Faraday declined, writing to a friend, "I must remain plain Michael Faraday to the last."

When Faraday heard of Oersted's discoveries, his interests shifted for a while from chemistry and electrolysis to physics. In order to learn the necessary experimental skills, Faraday reproduced *all* the known experiments involving electricity and magnetism, publishing his meticulous notes in the *Annals of Philosophy*. In total, Faraday performed over fifteen thousand experiments involving electricity and magnetism. While working on Oersted's results, Faraday invented the first crude electric motor, using electrical currents to move magnets, and by transforming electric energy into mechanical energy for the first time, pioneering the new field of electrical engineering. In 1823, despite strong opposition from a jealous Davy, Faraday, the son of the poor blacksmith from Surrey, was elected to the Royal Society.

Much more was to come. He found something very unpleasant about Oersted's discovery that electric currents generate magnetic forces. It was, in a way, too one-sided. What about the opposite? Can magnetic forces generate electric currents?

First, Faraday used his magnificent intuition to help him visualize the action of a charge on another through space. In his mind, there was no such thing as action at a distance. He imagined the influence caused by one electric charge on another, or one magnet on another, as a measurable disturbance in the space between them. That is, he imagined *lines of force* emanating from an electric charge or from a magnet, which would influence another charge or magnet placed at a distance from it. Faraday's imagined lines of force remind me of the words of the Fox in Saint-Exupéry's wonderful fable *The Little Prince*: "What is essential is invisible to the eye."

In Faraday's representation, the closer the lines of force are together, the stronger the effect of the electric or magnetic force. If you have two refrigerator magnets, you can play with them to feel the "lines of force" they create. Faraday's visualization technique is the precursor of the all-important concept of *field*.

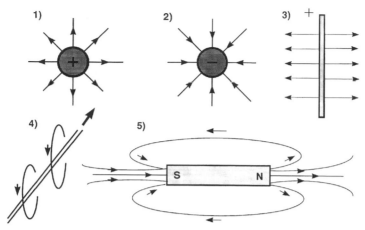

Examples of Faraday's field lines: 1) A positive charge. 2) A negative charge. 3) A portion of a very long, flat positively charged plate. The arrows indicate the direction of the force felt by a small positive charge placed nearby. 4) Magnetic field around a straight wire carrying a current. 5) Magnetic field of a bar magnet. The arrows indicate the direction of the force felt by the north pole of a small magnet placed nearby.

The presence of an electric charge disturbs the space around it in such a way that another charge placed nearby feels its presence through an electric force. The same is true of two magnets, or, for that matter, two masses attracted gravitationally. Thus, the electric field of an electrically charged body is measured by its action on other electrically charged bodies, which respond by either being attracted or repelled by the field. Every electrically charged body has an associated electric field. Every magnet has an associated magnetic field. And every mass has an associated gravitational field.

Faraday knew that the magnetic field created by an electric current flowing along a straight wire is shaped as concentric circles perpendicular to the wire. It was this circular magnetic field that was responsible for moving Oersted's compass needle. But how could a magnetic field generate an electric current? After several attempts, on August 29, 1831, Faraday was finally successful in generating the current. But there was a catch. In fact, this catch was the reason why it took so long for this effect to be discovered; the magnetic field had to be moving in time! This phenomenon is known as Faraday's *induction*. A constant magnetic field has no effect whatsoever.

A simple experiment can demonstrate this fact. Twist a wire into a wide cylindrical loop, and attach a galvanometer to the two ends of the wire. (A galvanometer is an instrument that can detect the passage of an electric current through the wire.) Now, move a magnet bar in and out of the loop. The galvanometer will indicate the passage of an electric current through the wire. Had you simply placed the magnet inside the loop and then looked at the galvanometer, nothing would happen. A moving magnet means a moving, or changing, magnetic field. A changing magnetic field creates an electric field that, in turn, induces the current in the wire. After all, a current is the *motion* of electric charges. The conclusion is simple, but of very deep significance: *Electricity and magnetism are unified through motion*. Electromagnetism was born. Faraday's belief in a deep unity in nature was vindicated:

> I have long held an opinion, almost amounting to conviction, in common I believe with many other lovers of natural knowledge, that the various forms under which the forces of matter are made manifest have one common origin; or, in other words, are so directly related and mutually dependent, that they are convertible, as it were, one into another, and possess equivalents of power in their action.

Faraday's conviction is a clear expression of the belief in that deeper layer of knowledge where apparent differences are understood as manifestations of common causes, or "origin." Electromagnetism makes much more sense than electricity and magnetism; perceiving the two as independent phenomena leads to an incomplete, fragmented description of the world. By peeling off one more layer of physical reality, Faraday revealed the hidden unity of electromagnetism in all its unparalleled beauty. It is no wonder that Faraday remains an icon to the many physicists searching for the hidden unity of nature.

Faraday's discovery of electromagnetic induction had major technological consequences: the dynamo, commonly used to convert power from a steam engine or a waterfall into electrical power;

the transformer, used to change the voltage of an AC current up or down for efficient power transmission; and the electric motor, capable of changing electricity into motion. Asked by the Chancellor of the Exchequer, "Of what use is all this?" Faraday replied, "One day, sir, you will tax it!" And indeed, Britain began taxing power usage about a half century later. Yet compelling as they were, something was missing in Faraday's breathtaking studies of electromagnetic phenomena.

By the mid 1850's a mountain of electromagnetic facts and laws obtained by Faraday and other physicists had accumulated, some written in terms of mathematical expressions, others just recorded as experimental results. With numerous experiments in search of a theory, some sort of organized way of presenting this wealth of information was badly needed. This is when James Clerk Maxwell (1831–1879), whom we just met in our discussion of heat, enters the scene.

The situation Maxwell encountered during his formative years has some parallels with the situation Newton encountered when he started his studies at Cambridge. Although Galileo had gathered a large amount of data and a few laws exploring motion and free fall, and Kepler had devised empirical laws describing planetary orbits, no synthesis existed putting all the pieces of the puzzle together. Newton not only integrated all the parts into a coherent whole, but also went much further, by building a solid conceptual foundation to the sciences of mechanics and gravitation. Maxwell did very much the same for electromagnetism. He put the pieces together into a coherent whole, then went much further. By establishing a solid conceptual foundation to the science of electromagnetism, he unveiled, as an unexpected bonus, the physical nature of light.

A math prodigy, Maxwell had submitted a paper to the Edinburgh Royal Society by the age of thirteen. Influenced by his father's love for tinkering with mechanical devices, he combined his mathematical ability with a keen intuition and dexterity in the laboratory. Maxwell's experimental talent made him appreciate the genius of Faraday, whom he revered throughout his scientific career. He became a Fellow of Trinity College at twenty-four, but soon left to chair the department of natural philosophy at Marischal College in Aberdeen. In 1857, Maxwell wrote a paper on the structure of

Saturn's rings that argued, correctly we now know, that the rings could remain in a stable orbit only if they consisted of minute particles. For this work he received the Adams prize, and a solid reputation. This work led to his interest in the motions of large groups of particles, which resulted in his ground-breaking discoveries in kinetic theory.

In 1860, Marischal College was absorbed by the University of Aberdeen, and Maxwell's job evaporated. He quickly joined the faculty at King's College, London, where he spent the next five years developing his electromagnetic theory.

Maxwell's first great accomplishment was to formulate the laws of electromagnetism in "local form." Faraday had discovered how lines of force spreading through space could account for action at a distance. If lines of force were a good representation of fields, then each point in space should have a field value attached to it. Or you may say that the field has a certain value at each point in space. This is what is meant by laws in local form (every point of space has a value). Say you are holding, by an insulating string, a positively charged little sphere as you approach a negatively charged body. The two attract each other, and do so more intensely as they get closer and closer. Hypothetically, if you don't hold on to your little sphere, it will be snatched away from you. (Just as when you are holding up a heavy object, fighting with the Earth's gravitational field!) The point is that you don't really have to know that there is a specific body attracting the little sphere. What you feel is the attraction caused by its *field*. For all practical purposes you can actually substitute its field for the large positively charged body.

There is a *very* big shift of emphasis on what constitutes the ultimate physical reality built into this formulation of electromagnetism. In Newtonian physics, we have particles and forces, and that's how the physical reality around us is described. But in come Maxwell and Faraday, and the focus moves on to fields, as opposed to particles, as the relevant objects to describe physical reality. Physics has never been the same since. As Einstein remarked in a speech celebrating the centenary of Maxwell's birth, "This change in the conception of reality is the most profound and fruitful one that has come to physics since Newton." Physical reality can be described locally by the values fields have in space, without explicit reference to their individual sources.

Maxwell organized all the experimental information then known about electromagnetic phenomena into four equations. But when he checked his equations, he noticed that something was wrong. Charge was not conserved! In order to "fix things up," he added an extra term, called the "displacement current," which explained not only how currents but also time variations in the value of an electric field could generate magnetic fields, similar to the changes in magnetic fields that Faraday found generated electric fields. A beautiful symmetry between the two fields was achieved. A time-changing electric field generated a magnetic field, and vice versa. This was his second great accomplishment. In the words of American physicist and Nobel prize winner Sheldon Glashow, "This small change in one of the four basic electromagnetic equations represented the greatest accomplishment in theoretical physics of the nineteenth century."

Maxwell's equations hid yet another jewel. Because they formulated how the electromagnetic field changed in space and time, the equations had built into them information on how *fast* the fields propagated in space. To Maxwell's surprise and delight, the computed speed for the propagation of electromagnetic disturbances in empty space was 186,000 miles per second. Electromagnetic fields moved at the speed of light!

The speed, or velocity, of a wave is given by the product of two numbers, its *wavelength*—the distance between two successive crests—and its *frequency*—the number of crests passing by a point in one second. It was well known by Faraday's time that Newton's corpuscular theory of light was incorrect. Light was better described by waves, albeit of very peculiar character.

If Maxwell's equations portrayed electromagnetic fields as waves moving at the speed of light, the conclusion was clear: The equations portrayed light itself in the form of an electromagnetic wave.

Let's pause to digest the enormity of this discovery. Take an electrically charged little sphere which has an electric field associated with it. Now start shaking it rhythmically up and down. As the charge is moved, its electric field changes in time. Since we know that a changing electric field gives rise to a changing magnetic field and vice versa, we know that a moving charge has an associated electromagnetic field. As the charge is moved up and down, its field oscillates. These oscillations propagate outward, just as concentric waves propagate outward from a pebble dropped on a lake.

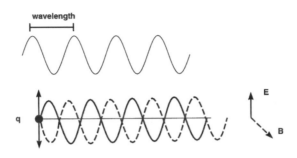

The wavelength is the distance between two successive crests. An oscillating charge (q) creates an oscillating electric field (E), which creates an oscillating magnetic field perpendicular to it (B). The result is a propagating electromagnetic field.

If the wavelength of these oscillations happens to be in the range of visible light, you can *see* the charge oscillating by the light emitted by it. In other words, light is created by accelerated charges. The kinetic energy of accelerating charges acts as the main source for the observed electromagnetic radiation, electromagnetic waves emitted by moving charges. There are also many forms of "invisible" *electromagnetic radiation*. They are just like light, but with wavelengths that our eyes cannot see. At wavelengths longer than visible light there are infrared radiation, radio waves, microwaves. At wavelengths shorter than visible light there are ultraviolet, X rays, and gamma rays.

The fact that we only "see" a very small band out of all possible electromagnetic radiation shows how limited our sensorial perception of the outside world is. But visible or invisible, electromagnetic radiation can be traced to accelerated charges. Nowadays, in an effort to improve our limited eyesight, astronomers "look" at the skies using

THE ELECTROMAGNETIC SPECTRUM

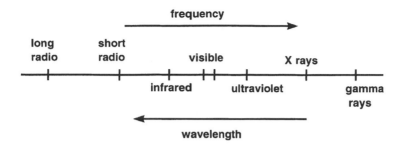

The electromagnetic spectrum: The visible portion makes for a very small part of the spectrum. The frequency of the waves increases to the right, while the wavelength increases to the left. For example, gamma rays are very high-frequency, small-wavelength electromagnetic waves.

all sorts of waves, from radio and infrared to X rays and gamma rays. And the views are breathtaking.

Maxwell's theory neatly synthesized many electric and magnetic phenomena. It predicted that light was an electromagnetic wave, and that there should be many other kinds of invisible electromagnetic radiation. But was his theory right? Unfortunately, Maxwell did not live to see its ultimate triumph. The lag time between theoretical prediction and experimental confirmation was beginning to increase.

Through a series of remarkable experiments starting in 1886, the German physicist Heinrich Hertz (1857–1894) was able to generate radio waves in the laboratory for the first time. He did so by showing how sparks generated in one circuit (the transmitter) could induce sparks to appear in another circuit (the receiver) placed five feet away. Guglielmo Marconi, with a keen sense for commercial enterprise, improved on Hertz's original demonstration by sending radio signals across distances of 10 meters, 30 meters, 3000 meters (a little under 2 miles), and then across the English Channel. In 1901, he sent the first telegraphic message across the Atlantic, the letter *s* in Morse code (dit-dit-dit), using radio waves with 200,000 vibrations per second (200,000 Hertz, or Hz for short) and a wavelength of al-most a mile.

There was a problem, though. A wave travels, or propagates, *on* something: a water wave on water, a sound wave on air. In fact, a bet-ter way to think of a wave is to say that it is the medium itself that waves, as a disturbance propagates through it. What about electro-magnetic waves? Through what medium can disturbances such as light or radio waves be perceived by us? Before Maxwell, it was be-lieved that a hypothetical medium, the *luminiferous aether*, or ether for short, had the sole function of supporting the propagation of light. Maxwell, who like everyone else at the time still believed in the existence of the ether, devised several contrived mechanical models to explain what the ether was. Using models made up as rods connected with balls and gyroscopes, he sought to illus-trate how the ether could support electromagnetic waves. Sounds bizarre? It was.

These attempts to construct mechanical models of the ether seem suspiciously like the Ptolemaic crusade to "save the phenom-ena," which resulted in the ever more elaborate models of the solar system complete with epicycles. The Greeks and their successors didn't believe that epicycles actually existed, but happily used (and abused) them as mental constructs to predict the locations of heav-enly bodies. Likewise, nineteenth-century physicists could not be-lieve they understood any phenomenon unless it was dressed in mechanistic Newtonian language. The fact that the ether had truly magical properties (like the Aristotelian fifth essence), filling all space yet being imponderable (weightless), being rigid as a solid yet never offering resistance to the motions of planets, didn't seem to bother them. But it should have. As we shall see, hidden in the mys-terious ether was a serious flaw of classical physics.

During the nineteenth century, classical physics went through a major expansion. The Newtonian mechanical worldview reigned supreme, celebrated by its countless successes. The description of motions both in the heavens and on earth, and many other exam-ples, confirmed the power of Newton's laws to describe natural phe-nomena. On the other hand, thermodynamics and the study of heat and electromagnetism and the study of light opened the door to a new physics with myriad possible technological applications. At the close of the century the world was a rapidly changing place, and the pace of change was only to increase.

In this busy world, even the God of the Deists was forgotten by most. Science became a profession and the study of nature a vocation devoid of any theological calling. In the biological sciences, Darwin dramatically increased the gulf between Church and science, by "dooming" humans to be direct relatives of the apes. The split between science and religion had become official and permanent. Scientific works were not to make any mention of the word *God*, focusing instead exclusively on the science. There was no need to justify the scientific enterprise by invoking divinity in nature, as Newton so eloquently did. Instead, we find a belief in the "unity" of natural phenomena, and in the beauty manifest in this unity, as the new fuel for science. Unity, beauty, and simplicity become the new words underlying the devotion to pure scientific research. Whether or not these words held any religious significance was left to the discretion of the practicing scientist.

By the end of the nineteenth century, so much was understood. The Sun and the stars were made of chemicals familiar to us here on Earth. Light emitted by distant nebulae was now seen as electromagnetic radiation produced by jiggling electric charges. The charges moved because they were hot, as the kinetic theory of Maxwell and Boltzmann explained. Thus, heat and light were united in the spectacular sights of the heavens. And yet so much remained unknown. If hot gases and objects emitted light because they had electric charges jiggling in them, what were these jiggling things? Why did different substances emit light of different colors? What was the ether on which electromagnetic waves propagated?

During the early decades of the twentieth century, physics would go through a major revision. Triggering this revision were explorations of these and other questions growing out of the discovery of electromagnetic radiation. From the heroic efforts of physicists dealing with these questions, relativity theory and quantum mechanics were born. With them a radically new interpretation of physical reality emerged, which deeply transformed our understanding of nature, from its smallest structures to the Universe itself.

PART IV

MODERN TIMES

7

OF THINGS FAST

The most beautiful experience we can have is the mysterious.
—ALBERT EINSTEIN

L earning modern physics can be an emotional experience. As students are first introduced to the ideas of relativity and quantum mechanics, initial perplexity is almost invariably followed by peevish disbelief. There is something outrageous about these theories, which seem to contradict our common sense. Among other things, the new physics tells us that a moving object contracts in the direction of its motion. A moving clock runs slower than a stationary one. Mass and energy may be interchangeable. The basic constituents of matter have the characteristics of both particles and waves, the so-called "wave-particle duality." The act of observing a physical system influences its behavior. The presence of matter can bend the geometry of space and alter the flow of time. We can't even say with certainty if something is here or there—only express the probability that it is here or there. That is, we must abandon a strictly deterministic description of natural phenomena, at least at the scales of molecules and atoms, and rely on probabilities. And there is more. Much more.

Common sense does not accord with these strange phenomena, and that's unfortunate, since we tend to rely on our common sense in the world around us. Webster's Dictionary defines common sense as "the unreflective opinions of ordinary people," or "sound and prudent but often unsophisticated judgment." Alternatively, we could say that common sense results from repetitive contact with certain situations, be they emotional or physical. Classical physics, to a large extent, deals with situations that are well within our direct sensorial experience. Even though certain basic results from classical physics, such as Galileo's observations of free fall and the law of inertia (Newton's first law), are somewhat counterintuitive (after all, even Aristotle was fooled), they deal with tangible situations; with some thought, it is not hard to realize that they make sense.

Not so with modern physics. Relativistic or quantum phenomena appear at first bizarre because they are far removed from our immediate sensorial reality; they are not part of the phenomena to which we can apply our "common sense." Indeed, the shrinking of moving objects or the slowing of time is manifest only at speeds close to the speed of light. The wave-particle duality of matter is relevant only for objects at the atomic scale. The effects of ordinary matter on the geometry of space and the flow of time are negligible for objects less massive than stars. Since we ordinarily deal with slow (compared to the speed of light), large (compared to an atom), and light (compared to a star) objects, our perception of the physical world is quite limited. Modern physics makes it clear that we should not have our expectations projected onto a realm that is alien to our everyday experiences. As Einstein has said: "Common sense is the set of all prejudices acquired by the age of eighteen." Once we look at relativistic and quantum phenomena with an open mind, what seemed nonsensical becomes fascinating.

Of course, it is easy for me to say this now, comfortably typing away on my computer, long after the dramatic discoveries of the first three decades of this century have been digested by several generations of physicists. But to the actual players in this drama, those thirty years were full of angst and despair. On several occasions, physicists were forced to put forward ideas or accept experimental results with which they felt very uncomfortable. Thus, Max Planck, the first man to propose that energy is emitted in discrete amounts (quanta), wrote of his discovery in an unpublished letter:

> Briefly put, I can describe the whole effort as an *act of desperation*, for by nature I am peaceful and against dubious adventures. But I

had been fighting already for six years, from 1894 on, with the problem of equilibrium between radiation and matter without having any success; I knew that this problem is of fundamental significance for physics . . . a theoretical explanation, therefore, *had* to be found at all cost, whatever the price. Classical physics was not sufficient, that was clear to me. . . . No matter what the circumstances, may it cost what it will, I had to bring about a positive result. [My italics]

In other words, Planck's quantum hypothesis was born out of a desperate attempt to understand the results of experiments that could not be explained classically.

Albert Michelson, whose brilliant 1887 experiment performed with Edward Morley was pivotal in establishing the nonexistence of the ether, could not accept his own results. What was supposed to be a routine confirmation of the existence of the celebrated medium turned into a nightmare. The ether was so dear to Michelson that to the end of his life he held on to it, despite the successes of Einstein's special theory of relativity, which elegantly proved how unnecessary this postulated medium was. Even as late as 1927, in his last publication, Michelson refers to the ether in nostalgic words: "Talking in terms of the beloved ether (which is now abandoned, though I personally still cling a little to it) . . ."

Change, for better or worse, always demands courage. Letting go of old, treasured ideas and the feeling of confidence they bring is hard to do. But as we look at the work of Galileo, Kepler, Newton, Faraday, Maxwell, Boltzmann, and many others we have encountered so far, it becomes clear that one of the most important characteristics of great scientists is freedom of thought. This independence brings with it a flexibility that allows them, with help from that elusive trait called genius, to find new and unexpected links where others see only dead ends. But finding new links is not enough; to chart new territory, scientists must also have the courage to let go of old, established notions. They must *believe* in their ideas.

While wrestling with his problems, Planck wrote, "I was ready for any sacrifice of my established physical convictions." When experiments exposed the limits of the accepted classical view, physicists were forced to take extreme positions in order to understand them.

Shocked by the "negative" results of Michelson's experiment, the great Dutch physicist Hendrik Lorentz wrote to Lord Rayleigh in 1892:

> I am totally at a loss to clear away this contradiction, and yet I believe that if we were to abandon Fresnel's theory [of the ether], we should have no adequate theory at all. . . . Can there be some point in the theory of Mr. Michelson's experiment which has as yet been overlooked?

A few years later, Lorentz indeed proposed a "fix" that could accommodate the results of the Michelson-Morley experiment. The fix boldly assumed that moving objects (including parts of the experimental apparatus) shrank in the direction of their motion. Ironically, Lorentz was right for the wrong reason. It was a clever hypothesis without a conceptual foundation to back it up. As Einstein later wrote in his landmark paper of 1905, "The introduction of a 'luminiferous ether' will prove to be superfluous." But Lorentz believed in it and put it forward nevertheless.

The final result of this tumult was a deep reformulation of the classical worldview. And as we have learned by now, whenever there is new physics, there is new cosmology; as our understanding of the physical world changes, so do our notions concerning the Universe as a whole. Thus, in order to understand the new models of the Universe proposed later in the twentieth century, we must first explore some of the revolutionary ideas that emerged as physicists wrestled with the world of the very fast and the very small. With this in mind, in the next two chapters we will plunge into the modern.

EINSTEIN IN COPACABANA

What is the only other name that could stand beside Newton in the gallery of science giants? We all have our biases, but I think that the other name must be Albert Einstein. Yes, I am an Einstein groupie. But then, so are most physicists. If you are not part of the fan club yet, I am sure that by the end of our discussion you will be applying for membership. (If I am wrong, it is entirely my fault, not Einstein's.)

Why does the name Einstein inspire such awe? Let's forget for a moment his popularized image as the wise old man with wild white hair, sweet dark eyes, and his tongue sticking out, a cross between an eccentric grandfather and a prophet. Einstein's scientific contributions stand unrivaled by any of his contemporaries in depth and diversity. Like Newton, Einstein developed a new conceptual foundation for physics, which deeply influenced the way his and subsequent generations of scientists, including my own, understand the world. Like Newton, his scientific contributions were not limited to a small corner of physics; he was a pioneer in several different areas. Unlike Newton, Einstein enjoyed discussing conceptual issues with his peers; his debates with the great Danish physicist Niels Bohr were crucial during the early development of quantum mechanics. Unlike Newton, he never got involved in bitter disputes concerning the originality of his ideas, or basked in the glory of his success. "The only way to escape the personal corruption of praise is to go on working," he said late in his life. "One is tempted to stop and listen to it. The only thing is to turn away and keep on working. Work. There is nothing else."

Einstein was much more than a scientist. He was a devoted pacifist who twice renounced his German citizenship due to his revulsion at its militarism. He continually voiced his outrage at a world shaken by two brutal wars. A tragic irony is that this man, who so often fought for world peace, fearing that the Germans would develop nuclear weapons, wrote to President Roosevelt in 1939, urging the United States to start researching the possible military uses of atomic energy. In 1954 he told the chemist Linus Pauling: "I made one great mistake in my life when I signed the letter to President Roosevelt recommending that atom bombs be made; but there was some justification—the danger that the Germans would make them." It is by now clear that the Manhattan Project would have been started with or without Einstein's endorsement.

Einstein's pacifism also found expression in his support for the Zionist cause. Even though his liberal views of the Jewish-Arab conflict often clashed with those held by most Zionist leaders, he eagerly lent his name and time to promote an independent Jewish state. The culmination of his involvement with Zionism was the invitation he received late in 1952 to succeed Chaim Weizmann as the president of Israel. Although a peculiar idea, the fact that Einstein was chosen to occupy the largely symbolic position gives a measure of his popularity. Einstein politely but firmly refused the invitation, saying to Prime Minister Abba Eban, "I know a little about nature, but hardly anything about man."

The media frenzy that erupted after the British astronomer Sir Arthur Eddington confirmed one of the predictions of the general theory of relativity in 1919 transformed Einstein almost overnight into a world-famous man. Much to his amazement and surprise, he became a public figure, a symbol of how a genius is supposed to look and act, the most famous scientist in the world, perhaps in history. The "otherworldly" nature of his ideas about space and time no doubt contributed to the creation of the myth, a point recently stressed by the physicist and Einstein's noted scientific biographer, Abraham Pais. Einstein seemed to have a direct pipeline to God, not unlike so many saints and prophets of the past.

I remember being fascinated by Einstein when I was still young. As soon as adults realized I enjoyed playing with chemistry sets and reading books about natural history, they would tell me of this great genius who had unlocked amazing secrets about the Universe. My father enjoyed summarizing Einstein's ideas in short sentences like "All is relative," or "Matter and energy are the same thing because $E = mc^2$." Plus, he was a Jewish scientist, and a Zionist to boot, which surely scored points with my family.

My fascination with Einstein grew to mythic proportions once I got his autographed picture from my step-grandmother. How Einstein's autographed picture ended up in an apartment in Copacabana is quite an interesting story. After Einstein became a public figure, he traveled around the world, visiting kings and presidents, and talking about his theory of relativity. Often he would participate in fund-raising activities sponsored by local Jewish communities trying to raise money for the Zionist cause. In May 1925, as part of a South American tour, Einstein came to Rio. The local Jewish community was of course very excited to meet its most famous world member. After much discussion, it was decided that Einstein would be hosted by both the Sephardic community (Jews of North African and Iberian origin), and by the Ashkenazi community (Jews of German and Eastern European origin).

One of the most active members of the Ashkenazi community

was Jacob Schneider, my maternal grandfather. The main representative from the Sephardic community was Isidoro Kohn, who was to be Einstein's main guide around town.* It seems that Einstein developed a fondness for Isidoro and his family. Before he left Rio, Einstein and Isidoro posed for a photograph, which they both signed afterward. As a token of his gratitude, Einstein insisted that Isidoro keep the tie he was wearing in the photograph. The photograph was carefully stored away in my step-grandmother's apartment until, when I was thirteen, she judged me the rightful heir to the prized relic. I could hardly believe my eyes. Despite the damage from decades of exposure to high tropical humidity, the signature was still legible, in surprisingly neat and rounded handwriting. I am still trying to get my hands on the tie.

You can imagine that to a dreamy teenager Einstein turned into an almost supernatural being. The more I learned about his work and thoughts, the more I realized that he indeed deserved all the praise, even though he didn't care much for it. But I also realized that there was more to Einstein than his scientific output (which was hard to understand at thirteen) or noble social concerns. What moved me then, and still does now, was his belief that science was an alternative path to organized religion as we confront the "mysterious."

In his autobiography Einstein wrote of his conversion, at about age twelve, from a deep religiousness into a faith in the redeeming power of science:

> When I was a fairly precocious young man I became thoroughly impressed with the futility of the hopes and strivings that chase most men restlessly through life. . . . As the first way out there was religion, which is implanted into every child by way of the traditional education-machine. Thus I came—though the child of entirely irreligious (Jewish) parents—to a deep religiousness, which, however, reached an abrupt end at the age of twelve. Through the reading of popular scientific books I soon reached the conviction that much in the stories of the Bible could not be true.

*It is an interesting twist of fate that when my father remarried in 1968, his wife was none other than Isidoro's niece, Léa. Thus, my life was twice linked to the great man, first through my mother and then through my stepmother. At least that's how my impressionable preadolescent mind saw it.

With his faith in organized religion shaken, Einstein found a focus for his intense need for spiritual freedom in the scientific study of nature:

> It is quite clear to me that the religious paradise of youth, which was thus lost, was a first attempt to free myself from the chains of the "merely personal," from an existence dominated by wishes, hopes, and primitive feelings. Out yonder was this huge world, which exists independently of us human beings and which stands before us like a great, eternal riddle, at least partially accessible to our inspection and thinking. The contemplation of this world beckoned as a liberation. . . . The mental grasp of this extra-personal world within the frame of our capabilities presented itself to my mind, half consciously, half unconsciously, as a supreme goal. . . . The road to this paradise was not as comfortable and alluring as the road to religious paradise; but it has shown itself reliable, and I have never regretted having chosen it.

Rarely have scientists written so passionately of their devotion to science. The "extra-personal" world stands as an eternal riddle, indifferent to humans but partially accessible to reason. The dedication to science was, to Einstein, the supreme goal, the path to self-transcendence. When I first encountered it, this view of science was different from anything I'd heard before. Einstein's words had the power of a magic spell.

Albert Einstein was born March 14, 1879, in Ulm, Germany. His parents, Hermann and Pauline Koch Einstein, were like most Bavarian Jews then: well assimilated and basically irreligious, although still clinging to certain traditions, such as marrying within the faith. Einstein was somewhat slow in his early development, starting to speak only after he was three, and apparently not completely fluent until nine. However, rather than an indication of a learning disability, it seems that Einstein's deficit was due to the fact that he was a very detached child, who was quite happy in his own little world. In fact, he never lost the ability to step easily in and out of immediate reality. As remarked by his biographer Pais, "He had no need to push the everyday world away from him. He just stepped out of it whenever he wished."

Another fallacy is that Einstein was a mediocre student. His grades were usually very high, quite often the highest in his class.* But he did deeply dislike the rigid authoritarian structure of German education. In fact, he disliked any form of authority, whether it is based in school, government, or religion. This may not have helped him win the support of his teachers, but without this courage to doubt, it is quite possible that much of his creativity would have been stifled. Einstein's ideas would not collect dust in the attic.

His romance with science started when he was five. He recalls in his autobiography the sense of wonder he felt when he was ill in bed, and his father showed him a compass: "I can still remember— or at least believe I can remember—that this experience made a deep and lasting impression upon me. Something deeply hidden had to be behind things."

His sense of wonder was strengthened at twelve, when he came across a little book on Euclidean geometry. Most impressive to him was the power of thought to prove complicated assertions on how curves, triangles, and circles intersected one another. From then on he took every opportunity to read books on mathematics and physics, with a seemingly endless appetite. And so it was that, when he was sixteen, a very precocious Einstein asked the question that would set him on the path to reformulate the Newtonian conception of absolute space and time.

Light Always Moves

Einstein later remembered the thought (or vision, really) that led him to the special theory of relativity this way:

> If I pursue a beam of light with the velocity c (velocity of light in a vacuum), I should observe such a beam of light as an electromagnetic field at rest though spatially oscillating. There seems to be no such thing, however, neither on the basis of experience nor according to Maxwell's equations.

---

*There is some confusion in the literature about this fact. I chose to follow Pais's data. But if Einstein's grades were weak, it would only serve to prove the inefficacy of a rigid educational system to accommodate genius.

This situation seemed to the young Einstein quite paradoxical. After all, according to Newtonian physics, if a wave is moving forward with some speed, we just have to move a little faster than it does to eventually catch up with it. And once we do, as every surfer knows, the wave will not be moving forward with respect to us anymore. The same should be true for an electromagnetic wave, since in Newtonian physics there is nothing special about the speed of light, apart from the fact that it is very high. But according to Maxwell's theory, there is no such thing as an electromagnetic field at rest in empty space. Light is always in motion. Something had to give: either Newtonian ideas concerning relative motion or Maxwell's theory of electromagnetic fields. What finally gave was the idea that the velocity of light was just like any other velocity.

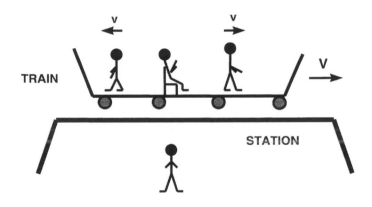

An observer watches a train moving eastward with velocity V. Inside the train, passengers are sitting down and moving both eastward and westward with velocity v.

Let's think about this in more detail. Consider a train moving eastward ($\rightarrow$) with constant velocity V with respect to an observer standing at a station. The first thing to note is that, to a passenger sitting in the train, the station is moving westward ($\leftarrow$). When we say something is moving, it is always in reference to something else not moving with it, be it ourselves, a tree, or a train station. In other words, motion is always in relation to some reference. Now imagine the following situation (a thought experiment): A passenger in the train is walking to the restaurant coach with velocity v eastward ($\rightarrow$) with respect to the passenger sitting in the train.

It is clear that to the person standing at the station, the walking passenger is moving eastward with velocity $V + v$ ($\rightarrow$). It is also clear that if the walking passenger was moving westward ($\leftarrow$), the person standing at the station would measure her velocity as $V - v$. This is all perfectly fine according to our common sense and Newtonian mechanics. In addition, the motion of the walking passenger can be equally gauged by the passenger sitting in the train, or by the person standing at the station. This result is encapsulated in the principle of relativity, which states that the laws of physics are identical for observers moving at constant velocity with respect to each other. For example, energy is equally conserved for both observers. If they know their relative speeds, they can compare results and agree with each other (to high accuracy, but not exactly, as we will soon see). The train and the station are called inertial reference frames. Reference frame here refers to a place in space from which an observer can make measurements of position and time. Inertial refers to the fact that this reference frame is either at rest or moving with constant velocity with respect to another inertial frame. For example, a car moving at constant velocity with respect to an observer standing by, or the train and the station. The principle of relativity states that the laws of physics are identical for inertial reference frames. For noninertial reference frames, such as a train accelerating with respect to the station, we need a more complicated theory, Einstein's general theory of relativity.

Now comes the interesting part. Instead of a walking passenger moving with velocity v to the restaurant coach, imagine that the sitting passenger stands up, points a flashlight toward the east ($\rightarrow$), and turns it on. "No problem," you say, "light coming out of the flashlight will move eastward with velocity c (the notation for the speed of light) with respect to the train, and with velocity $V + c$ with respect to the person standing at the station. Right?" Wrong! If this were true, you could imagine a situation in which the passenger would point his flashlight westward ($\leftarrow$), and if the train's velocity eastward ($\rightarrow$) was equal to the velocity of light, the person standing at the platform would see a beam of light at rest, contradicting Maxwell's theory that an electromagnetic field cannot be at rest in empty space. How, then, could Maxwell's theory be reconciled with the principle of relativity?

As a way out, Einstein suggested that the velocity of light in a vacuum (empty space) is not like any other velocity, but is special. It is the limiting velocity for causal processes in nature, that is, the fastest velocity anything can travel. More than that, it is *independent* of the velocity of its source. To the passenger holding the flashlight, the velocity of light waves emanating from it is c; it is the same velocity for the person standing at the station. With this proviso, Maxwell's theory can be made compatible with the principle of relativity.

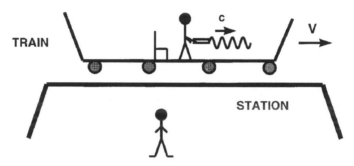

An observer watches a train moving eastward with velocity V. Inside the train, a passenger points a flashlight to the east and turns it on. Light coming out of the flashlight will travel east with velocity c for *both* observers.

In 1905, Einstein wrote a remarkable paper entitled "On the Electrodynamics of Moving Bodies," in which he elaborated his idea that the speed of light is unique. He had finally solved the problem he had posed to himself ten years before. He was then a "third-level" patent officer in Bern, Switzerland, after searching unsuccessfully for an academic position. Einstein's continual absence from lectures and his dedication to his own interests had not made him popular with his teachers. But we shouldn't feel sorry for him. He always referred to his days at the patent office in Bern as the happiest in his life. He had plenty of time for his research in physics, and his personal life was also quite felicitous. In 1903 he had married Mileva Maric, a former classmate at the Polytechnic Institute (ETH) in Zürich. A year later they had the second of their three children, a boy named Hans Albert.*

*While students at the ETH, Mileva and Einstein had a daughter who was sent to Mileva's parents while very young and was never heard of again.

In his ground-breaking paper Einstein developed the foundation of the special theory of relativity from two postulates:

1. The laws of physics are the same for observers moving with constant velocities with respect to each other.
2. The velocity of light in empty space is independent of the motion of its source and the motion of the observer.

The first postulate is the old principle of relativity. Physics is the same for inertial reference frames. Observers can meaningfully compare results with each other. But the second postulate, that light always moves with the same speed, was new. Innocent as this postulate sounds, it had profound consequences for several Newtonian notions of space and time. Einstein's brilliant insight was to insist that the principle of relativity be consistent with the constancy of the speed of light. In fact, by enforcing these two postulates together, Einstein is making sure that the constancy of the speed of light should be the same for all inertial observers.

TRAINS, CLOCKS, AND STICKS

In order to appreciate some of the amazing consequences of Einstein's special theory of relativity, we must agree on the definition of an *event*. An event is something that happens, an occurrence. It takes place at some location in space and at some moment in time, as for example, a ball hitting the ground. Note that in order for an observer to make a measurement of the position and time of an event, he must "see" what is going on. That is, information must be transferred between the event and the observer. Light, and its finite speed, become part of the measuring process. Einstein's second postulate leads to the following surprising result: *Simultaneity is relative.* Two events that are simultaneous for observer A, such as two balls hitting the ground at the same time, will not be simultaneous for observer B moving with constant velocity with respect to observer A.

You don't believe me? Okay, let's go back to the moving train and observer A standing at the station.* As before, the train is moving eastward ($\rightarrow$) with velocity V with respect to A. Sitting right in

*The arguments below are inspired by Einstein's own popular account of relativity, which is a model of clarity.

the middle of the train is observer B. Suddenly, something frightening happens. Observer A sees two thunderbolts strike the front and back of the train at exactly the same time. (Don't worry, no one gets hurt in a thought experiment.) Observer A knows they struck at the same time because light from each of the thunderbolts took exactly the same time to travel toward their midpoint, M. The two events are simultaneous for A. But are they simultaneous for B, who is moving eastward with velocity V? Since B is riding toward the light from the thunderbolt that hit the front of the train, and away from the one that hit the back, she will see the light from the thunderbolt hitting the front of the train *before* she sees the light from the thunderbolt hitting the back. Thus, for B the two events are not simultaneous.

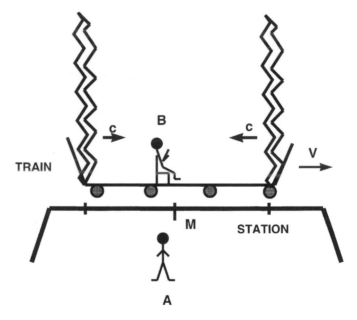

Observer A watches two thunderbolts strike the front and back of a train moving eastward with velocity V. A concludes they were simultaneous events, as light took the same time to reach their midpoint M. A second observer B moving with the train concludes otherwise, as light from the thunderbolts reaches B at different times.

What is simultaneous for one observer is not for another. Every observer has his or her own particular time, only correlated if they know their relative state of motion, or relative velocity. There is no such thing as absolute time.

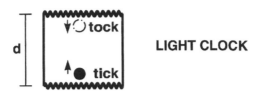

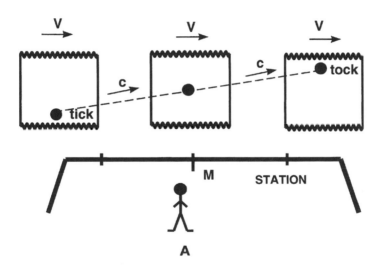

In our thought experiment on time dilation, a light clock with height d "ticks" when the light pulse hits the bottom mirror and "tocks" when it hits the top mirror. For an observer at rest with the clock, the time interval between a tick and a tock is t_0. However, for the same observer, as the clock moves it will tick slower, since the light pulse must travel a larger distance than d to reach the top mirror. In this case, the time interval between a tick and a tock is t_v, larger than t_0.

Two other consequences of Einstein's second postulate defy common sense. They are known as *time dilation* and *length contraction*, respectively. Basically, they state that (1) a moving clock ticks slower than one at rest; and (2) a moving object shrinks in the direction of its motion. If the clock and the object achieve the speed of light, time stops and the object shrinks to nothing. Perplexed? Actually, these ideas are not as strange as they seem. I will first explain to you why moving clocks run slower.

Let's go back to the train, which is now at rest at the station. First, put a device in the train, which goes by the name of "light clock" (see figure on previous page). It consists of a transparent box of height d that has two identical mirrors, one on the floor and one on the ceiling, facing each other. Somehow (this is a thought experiment!) it is possible to make a light pulse inside the box, which then bounces up and down like a rubber ball as it is reflected by the mirrors. Whenever the light pulse hits the bottom mirror, we hear a "tick" sound, and when it hits the top, we hear a "tock" sound. Before the light clock was put in the train, observer A in the station measured the time interval between a tick and a tock, and called it t_0. Now, the train backs up a certain distance and starts moving toward the station, eventually passing A with constant velocity V. Observer A hears a "tick," followed by a "tock." She calls the time interval between the two t_V. When she compares the two measurements, she notices that t_V is larger than t_0: The time interval between a "tick" and a "tock" increased for the moving clock!

Once observer A recovers from her initial shock, she concludes that it couldn't have been otherwise. Let me reconstruct her reasoning. For observer A standing by the station, the path the light pulse took to go from the bottom mirror to the top was longer for the moving train when the light clock itself was moving, and the pulse had to travel farther to hit the top mirror. That is, the light pulse traveled a distance larger than d. *Since light always travels at the same speed* (second postulate), and speed is distance over time, A concludes that, when moving, clocks tick slower and thus are slower! Notice that this effect is only for observer A. Observer B in the train, who is moving with the clock, would measure exactly t_0 for the time interval between the "tick" and the "tock," just as A had when the clock was at the station.

This result has nothing to do with the particular clock we used. Had we used the beating of a heart as our clock, the results would have been the same. When in motion, biological time or any other time slows down.

Finally, Einstein's postulates introduce the notion of length contraction. Let's repeat the "light clock" experiment, but now with the clock lying horizontally, as shown in the figure below. Observer A at the station measures the time between a "tick" and a "tock" for the moving clock as before. Since the orientation of a clock cannot affect its performance, A measures the same time interval as before, t_v. However, now the light pulse has to cover not only the distance from one mirror to another, but also it has to catch up with the receding mirror. *Since light always travels at the same speed,* the only way that the times can be the same is if the distance between the two mirrors shrank. Moving objects contract in the direction of their motion!

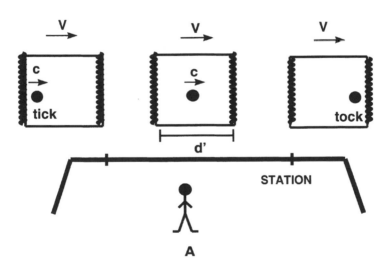

Length contraction: The light clock is laid over lengthwise in the direction of its motion eastward. Now the light pulse has to catch up with the receding mirror. Since observer A measures the same time as when the clock is standing up straight, she concludes its length must be contracted to d'.

"Wait a minute," you say. "If Einstein is right, why do we never observe the relativity of simultaneity, shrinking moving objects, or slow-
moving clocks?" The reason is that the speed of light is so much
higher than the ordinary speeds of everyday life that for us these effects are negligible. For example, for a moving clock to slow down by
40%, it would have to be moving at 80% of the speed of light, or
about 150,000 miles per second! That's why Newtonian physics works
so well for us. A slow-moving world is very well described by classical
physics. "But again," you would insist, "if these effects are negligible
in everyday life, how do I know if Einstein's ideas are right? And why
should I care?" These are both very good questions.

We know Einstein's special theory of relativity is right because
even though we can't move close to the speed of light, other things
in nature can. Some of these fast-moving objects can be found in
cosmic rays, showers of tiny bits of matter that keep pouring from
outer space into our atmosphere. As you read these lines, you are
being bombarded by these particles. Although we don't know for
sure where they come from, we suspect cosmic rays originate from
violent astrophysical events such as supernova explosions, which
mark the death of very massive stars. When cosmic rays (mostly protons) strike atoms in the upper atmosphere, they produce, among
other debris, a particle called a *muon*, a heavy cousin of the electron.
We know from laboratory measurements that the muon is an unstable particle with a half-life (that is, the time it takes half of the atoms
or particles in a sample of material to decay, or disintegrate) of two
millionths of a second; after this time, muons (on average) decay
into other particles. If the muons travel near the speed of light,
Newtonian physics tells us that they would cover about 600 meters
(a little under 2,000 feet) before decaying. But physicists detect
them hitting the ground after traveling well over 50,000 feet! Special
relativity easily explains that: If the muons travel at 99% the speed of
light, they can cover 12,000 feet before decaying. Cosmic ray muons
can travel even faster, explaining why we observe them at ground
level. Fast-moving muons live much longer than slow-moving ones.

There are many other situations where we observe the effects of
time dilation and length contraction in precise agreement with
special relativity. Particles whiz around at speeds close to the speed
of light every day in big underground machines called particle

accelerators, like the one at Fermilab, forty miles west of Chicago. Some particles exist for such short times that we would have no way of observing them if not for time-dilation effects.

This brings me to the second question: Why should we care? There are several reasons. The most obvious one is that if we were to rely on our limited sensorial perception of physical phenomena to develop a comprehensive understanding of nature, the final result would be a very incomplete picture. Science expands our senses into invisible worlds that defy our imagination, be they cells, atoms, or distant galaxies. This is probably why some people choose to become scientists in the first place. Once in a while they stumble onto something new, a previously invisible world that invariably expands our intellectual horizons and sometimes will be of practical use to humankind. Without knowledge of the cell or the atom, much of our medical and technological advances would simply not exist. It is very hard to think of life without penicillin or computers.

Unfortunately, there are two sides to invention. As Siddhartha Gautama (the Buddha) said, "Wherever there is light, there is shadow." Knowledge can bring power, and power is seductive. Science can cure, but it can also kill. But while this is true, the alternative is certainly not to deny the crucial relevance of science to society. That would be a guaranteed one-way ticket to a backward society, forcing the standard of living to regress to the miserable levels of the not so distant past. Not that misery is not present in the world today, far from it. But at least the possibility of alleviating it through the benefits of modern science exists. Knowledge may not always mean wisdom. But ignorance is certainly never a reasonable option.

Let's now return to the Michelson-Morley experiment, which failed to prove the existence of the ether, where our discussion of twentieth-century physics began. I introduced Einstein's ideas without referring to the famous experiment because evidence indicates that Einstein was not aware of the results when he was formulating the conceptual foundation of relativity. Or if he was, the experiment did not play a relevant role in his thoughts. What motivated Einstein was the incompatibility of the principle of relativity with Maxwell's electromagnetism. This is where Einstein's position differed from that of the Dutch physicist Hendrik Lorentz, and before him, the Irish physicist George Fitzgerald (1851–1901), who had proposed length contraction to keep the existence of the ether consistent with the "negative" results of the Michelson-Morley experiment. But while they wanted to save the ether at all costs, even if it meant conjecturing a bizarre contraction of objects in the direction of their motion, their proposal lacked a firm conceptual foundation. Instead, to Einstein, the ether was completely unnecessary. The length contraction postulated by Lorentz and Fitzgerald follows automatically once it is imposed that the speed of light never varies under the principle of relativity. This was not merely a quick fix, but a deep conceptual shift. In the words of Gerald Holton:

> Lorentz's work can be seen somewhat as that of a valiant and extraordinary captain rescuing a patched ship that is being battered against the rocks of experimental fact, whereas Einstein's work, far from being a direct theoretical response to unexpected experimental results, is a creative act of disenchantment with the mode of transportation itself—an escape to a rather different vehicle altogether.

Einstein's special relativity established how observers in relative motion (with relative constant velocity) could meaningfully compare their measurements. After a beautiful mathematical reformulation developed by the German mathematician Hermann Minkowski in 1909, it became clear that special relativity required space and time to be related in such a way that it is best to think of them as jointly forming a new four-dimensional reality, called spacetime, where time and space are treated in equal footing. Thus, a "distance" in spacetime encompasses both space distances and time intervals. The two postulates of special relativity guarantee that distances in spacetime are preserved under relative motion. In a sense, relativity is a misnomer, as the theory is built upon quantities that remain the same for all inertial observers. The strange effects such as time dilation and length contraction are only caused by looking at physical reality through the myopic lenses of the sensorial space and time of Newtonian physics. The best arena to describe physical events is the four-dimensional spacetime of special relativity, where distances are the same for all inertial observers. The special theory of relativity is in fact a theory of absolutes.

The three consequences of special relativity discussed so far, the relativity of simultaneity, length contraction, and time dilation, are complemented by another one, presented by Einstein in a second paper also published in 1905. In the celebrated $E = mc^2$ formula, mass is a form of energy. Even if an object is not moving, it has energy stored in its "rest" mass, m. But what if the object is moving? It should have more energy than when it is at rest. In order to accommodate this obvious fact, Einstein proposed that an object's mass increases with speed, reaching an infinite value as the speed of light is approached. But since it would take an infinite amount of energy to accelerate an object to the speed of light, no object with spatial extension and mass can reach the speed of light. It is, contrary to the claims of science fiction, the ultimate speed in nature. Only the poetic imagination of a precocious teenager could travel as fast.

8

OF THINGS SMALL

*Heaven knows what seeming nonsense may not tomorrow
be demonstrated truth.*
— ALFRED NORTH WHITEHEAD (1925)

L ight, or more generally, electromagnetic waves, posed other
challenges for classical physics. We have seen that light emitted and absorbed by chemical elements and analyzed by a spectroscope allowed physicists to study the chemical composition of the
Sun and distant nebulae. However, as the nineteenth century drew
to a close, no one knew why different elements have different spectra, or why spectra existed in the first place. To make things worse,
there was no proper explanation as to why certain objects, such as
iron pokers or lamp filaments, glow different colors when they are
heated to different temperatures. In the end, the answer to this de-
ceptively simple question held the key to a profound revolution in
physics. It is a story worth telling, not only because of the wide
repercussions of quantum physics to our understanding of the Universe big and small, but also because it serves as an excellent example of how progress in physics often advances in tortuous ways.

THE COLOR OF HEAT

We all know that if we heat up an iron poker to sufficiently high
temperatures, say in our grandmother's fireplace, it will eventually

start glowing in a reddish tone. If your grandma gets a really powerful fire going, increasing the temperature will make the poker glow yellower, and at even higher temperatures, the poker will emit a bluish glow. (Actually, this will depend on the type of material, since iron melts before turning blue.) An electric range is also a great laboratory for seeing how hot things glow. As you crank up the dial, the invisible heat (infrared radiation) emanating from the metallic spiral becomes visible, gradually changing from a faint to a very strong orange-red glow. Classical physicists could understand this behavior by marrying thermodynamics with Maxwell's electromagnetism. If the poker is made of electric charges that somehow vibrate (there was no model of the atom yet!), then, as it gets hotter the charges will vibrate faster, emitting radiation of higher frequency. Since blue has higher frequency than red, the hotter the poker, the bluer the glow. So far so good.

As more detailed questions were asked, though, classical physics began to flounder. Soon it became clear that new ideas were desperately needed, but no one had a clue where to start. As unexpectedly as the Michelson-Morley experiment, the familiar hot-red poker had turned into a nightmare.

Consider the first obstacle to be faced: Objects made of different materials and of different shapes have different thermal properties. During the late 1850s, about the same time he was investigating the chemical composition of the Sun (another hot body that glows!), the German physicist Gustav Kirchhoff suggested a method physicists could use to study the radiation emitted from a hot body without worrying about its composition, shape, or size. Little did he know that a deep conceptual revolution underlay his clever idea.

The object Kirchhoff suggested was a closed cavity, like the interior of an oven or a kiln, which he would then heat to some temperature T. Since heat induces motion, the molecules making up the walls of the cavity would move about, collide, and emit electromagnetic radiation into the cavity. In turn, the electromagnetic radiation in the cavity would be reabsorbed by the walls, and a dance of equilibrium between emission and absorption would quickly be established. Kirchhoff showed that since emission and absorption "cancelled each other out," the spectrum inside the cavity would include no spectral lines (all chemical signatures were erased), and thus it could not depend on the shape, size, or material the cavity was made of. Since a perfectly absorbing surface is black, while a perfectly reflecting surface is white, Kirchhoff's cavity, which absorbed all the heat it received without emitting any, became known as a *blackbody*.

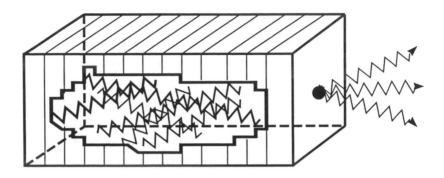

In an oven maintained at temperature T, blackbody radiation escapes through a small hole in one of the walls.

In order to study what kind of radiation was inside the cavity, Kirchhoff made a tiny hole to allow some of it to leak out. The resulting spectrum, called blackbody spectrum, displays electromagnetic radiation of all frequencies, each carrying a certain amount of energy with it. In a rough sense, the interior of the cavity is like a choppy ocean with superimposed waves of all sizes. Since the temperature is the only parameter characterizing a blackbody, the amount of energy carried by radiation of a given frequency is determined solely by the temperature. Every object radiates—from a tungsten filament in a light bulb (visible) to human bodies (invisible infrared)—producing a spectrum that can, with better or worse results, be approximated by a blackbody.

The fact that temperature alone determines how much energy will be carried away in each frequency of the blackbody spectrum is just the kind of universal behavior physicists like. As German physicist Max Planck wrote in his *Scientific Autobiography*: "[Kirchhoff's result] represents something absolute, and since I had always regarded the search for the absolute as the loftiest goal of all scientific activity, I eagerly set to work."

The search for the absolute is a persistent muse for scientific creativity, an inspiration stretching all the way back to the pre-Socratics. What Planck was after was a theory that could explain the exact dependence of the blackbody spectrum on temperature. That is, given the temperature, how much energy was emitted in a certain frequency of yellow, how much in a frequency belonging to blue, and so on. Experimentalists measured that the power (energy per second) emitted by the blackbody increased with frequency, reaching a maximum and then dropping for even higher frequencies. They also found that the frequency of maximum brightness changed with temperature, moving from red to blue as temperature increased. The task at hand was to find a simple mathematical relationship that could explain these results, using thermodynamics and Maxwell's electromagnetism.

However, when classical physics was applied to the blackbody problem, it predicted a spectrum very different from the one measured in the laboratory. Instead of finding that the power (or brightness) emitted increased with frequency and then decreased after reaching a maximum, classical physics predicted that the power emitted always increased with the frequency. Roughly speaking, it predicted that the red-hot poker should be blue.

Desperate to find a solution to the problem, on October 19, 1900, Planck announced to the Berlin Physical Society that he had obtained a formula that nicely fit the results of the experiments. But a formula was not enough. In order to truly understand the physics behind a phenomenon, more than a fit, or fix, is needed. Planck was fully aware of this fact. He later wrote, "On the very day when I formulated this law, I began to devote myself to the task of investing it with a true physical meaning."

In order to uncover the physics behind his formula, Planck was led to the radical assumption that atoms do not give radiation away in a continuously rising flow, but in discrete multiples of a fundamental amount. Thus, atoms dole out energy the way we dole out money, in multiples of the smallest unit of "currency." For each frequency there is a minimum discrete "cent" of energy, which is proportional to the frequency; the higher the frequency, the larger the "cent."* Radiation

*The mathematical expression for the minimum "cent" of energy is $E = h v$, where v is the frequency of the radiation, and h is a constant known today as Planck's constant. Planck originally obtained its value by fitting his formula to the experimental results.

of a particular frequency can only appear as multiples of its fundamental "cent," later called *quantum* by Planck, a word that in late
Latin meant a portion of something. As the great Russian-American
physicist George Gamow once remarked, Planck's hypothesis of the
quantum created a world in which you could drink a pint of beer or
no beer at all, but nothing in between.

Planck was far from happy with the consequences of his quantum hypothesis. In fact, he spent years trying to "explain" the existence of a quantum of energy using classical physics. He was
a reluctant revolutionary, led by a deep sense of scientific honesty
to propose an idea he was not comfortable with. As he wrote in his
autobiography,

> My futile attempts to fit the . . . quantum . . . somehow into the clas-
> sical theory continued for a number of years, and they cost me a
> great deal of effort. Many of my colleagues saw in this something
> bordering on a tragedy. But I feel differently about it. . . . I now
> knew that the . . . quantum . . . played a far more significant part in
> physics than I had originally been inclined to suspect, and this
> recognition made me see clearly the need for the introduction of
> totally new methods of analysis and reasoning in the treatment of
> atomic problems.

Planck was right. The quantum theory he helped propose
evolved into an even deeper departure from the "old" physics than
Einstein's special relativity. Classical physics is based on continuous
processes, such as planets orbiting the Sun, or waves propagating on
water. Our whole perception of the world is based on phenomena
that continuously evolve in space and time. But the world of the very
small is completely different. It is a world of discontinuous processes, a world where rules alien to our everyday experience dictate
bizarre forms of behavior. However, we are effectively blind to the
radically different nature of the quantum world. The same way that
the slowness of our everyday speeds precludes us from observing the
effects of special relativity, the energies we commonly deal with con-
tain such an enormous number of energy quanta, that their discreteness is effectively lost to us. It is as if we lived in a world of

billionaires, where a cent would be a perfectly negligible amount of money. But in the world of the very small, the quantum rules.

Planck's initial reluctance to push forward the quantum hypothesis was in direct contrast to ideas coming from the patent office in Bern. Again in 1905, the same year he wrote his papers on the special theory of relativity, Einstein wrote two more papers, each brilliant enough to immortalize him. One addressed the interesting phenomenon known as Brownian motion, where small grains (for example, pollen) when floating on a liquid exhibit an erratic zigzag motion. The British botanist Robert Brown had discovered this behavior in 1827, while observing grains of pollen floating on tiny drops of water under a microscope. Initially he thought the motion was caused by some occult life force in the grains of pollen. However, he later showed that any fine particle, organic or inorganic, experiences the same erratic motion when suspended in liquids. Einstein (and independently, the Polish physicist Marian Smoluchowski [1872–1917]) showed that the erratic motion was due to the collisions of the small grains with the molecules of the liquid, thus offering strong support for the atomic nature of matter, assumed by Boltzmann in his formulation of statistical mechanics.

Einstein's fourth great paper of 1905 (actually published first) was on the subject of the so-called *photoelectric effect* discovered by Hertz in 1887. He found that certain types of electromagnetic radiation striking an electrically neutral piece of metal could make it gain a net positive charge. It was a curious effect that baffled physicists who tried approaching it with Maxwell's electromagnetic theory. For example, no one could understand why yellow light would not electrify the metal, whereas blue (or ultraviolet) would. It was clear that the effect could be explained if somehow light struck off electrons from the surface of the metal. Since electrons carry a negative electric charge, a metal depleted of electrons would become positively charged. But Maxwell's theory could not explain why yellow light didn't work while blue did.

In order to solve the problem, Einstein proposed to extend Planck's hypothesis that atoms radiate in discrete packets to *light itself*! It was a very bold move. As the historian of science I. Bernard

Cohen wrote, "Chiefly it was his [Einstein's] paper of March 1905 which marks the transformation of Planck's potentially revolutionary idea into a truly revolutionary one."

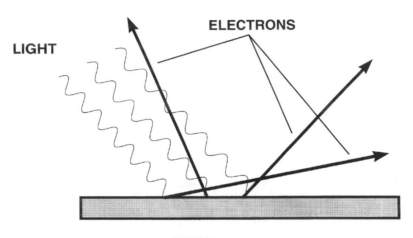

The photoelectric effect. When light of sufficiently high frequency hits a neutral piece of metal, it removes electrons, giving the metal a net positive charge.

As with Planck's energy quanta, Einstein suggested that light of a given frequency comes in multiples of little packets, each with energy proportional to the frequency ($E = h\nu$). Einstein pictured the radiation inside the blackbody cavity as a "gas" of light particles with energies proportional to the several frequencies of the radiation. A crude analogy would be to think of the radiation inside the cavity as a gas made of billiard balls of different "colors" (visible and invisible!), one for each frequency. Atoms from the cavity's walls could "eat" billiard balls of the same frequency, and spit them back into the cavity, but not half balls. Einstein effectively extended Boltzmann's atomistic treatment of matter to light. These "atoms of light" were called photons in 1926 by the American physicist Gilbert Lewis. Once this atomistic hypothesis for light is accepted, the mystery of the photoelectric effect quickly fades away. Yellow light couldn't charge the plate because it is of lower frequency (and thus energy) than blue light. It just didn't have enough energy to knock off electrons from the metallic surface.

"Wait a minute," you exclaim with an irritated tone in your voice, "didn't Maxwell and countless experiments show that light, or electromagnetic radiation of any frequency, is a wave? Are you trying to confuse me?" I am not, I promise. Einstein, of course, knew this as well as anyone else, and simply suggested his light quantum as a "heuristic" hypothesis, meaning an exploratory one of temporary validity.* In other words, he didn't know why it worked, but he knew it did. Well, you can imagine that if scientists were already struggling with his special relativity, Einstein's new quantum hypothesis for light was not received with pats on the back. Was light a wave or a particle? And who was this fellow from the patent office in Bern anyway?

Einstein didn't seem to care. He was quite happy with his promotion in April 1906 to technical expert second-class, which came with a nice raise in salary. By 1908, however, Einstein had decided to move on with his academic career. He obtained a *venia docendi*, the right to teach, and looked for a part-time teaching job that would allow him to keep his secure position at the patent office. A measure of the reluctance of the academic community to accept Einstein's revolutionary ideas is that only in 1909 did he receive his first job offer as a regular faculty member. In fact, the initial reactions to special relativity were experimental results that (erroneously) tried to refute it. Not until 1915 would Einstein's detailed predictions concerning the photoelectric effect be proven beyond doubt, although very reluctantly, by the American physicist Robert Millikan. In 1948, he recalled:

> I spent ten years of my life testing that 1905 equation of Einstein's, and contrary to all my expectations, I was compelled in 1915 to assert its unambiguous verification in spite of its unreasonableness, since it seemed to violate everything we knew about the interference of light [a wavelike property].

---

*This choice of word, used by Einstein in the title of the paper on the photoelectric effect, expressed his sentiment, which he would not abandon to the end of his life, that the quantum theory is a provisional theory, a sentiment mirrored by Planck's initial attempts to explain his constant classically.

In spite of this initial inertia, Einstein's brilliance was recognized by a few influential physicists, including Planck and Lorentz, and his reputation slowly spread throughout Europe.

When Einstein was being considered for the position of associate professor of physics at the University of Zürich in 1909, the proposal to the faculty stated: "Today Einstein ranks among the most important theoretical physicists and has been recognized rather generally as such since his work on the relativity principle." From then on, his reputation skyrocketed almost as fast as the speed of light. During the same year he received an honorary doctorate from the University of Geneva, together with Marie Curie and Wilhelm Ostwald. In 1910, he moved to Prague for a full professorship and a higher salary, returning to Zürich after two years, only to finally settle in Berlin in 1914 as director of the prestigious Kaiser Wilhelm Institute. Although Einstein did not enjoy living in Germany, the temptation of being at a top research institute close to luminaries like Planck and the physicist Walter Nernst, with a much higher salary and few teaching duties, was hard to resist. His wife, Mileva, did not share his excitement, and moved back to Switzerland with their two sons.* Their marriage was falling apart.

The Quantum Waltz

While physicists were struggling with the disturbing new concepts of the special theory of relativity and the quantum hypothesis, a barrage of new and bizarre experimental results was mercilessly accumulating. In the relatively short period of sixteen years, physicists went from not having much of an idea if atoms existed or not, to the discovery of X rays, radioactivity, electrons, and the atomic nucleus. What incredible times for physics these were.

First came Wilhelm Röntgen's discovery of X rays in 1895. Like Fraunhoffer's discovery of the solar spectrum, it was one of those episodes in which chance helps the well prepared. In those days

---

*Their second son, Eduard, had been born on July 28, 1910. He was a sensitive and melancholy child, who showed a talent for the arts that, sadly, could never really unfold. As reported in Paris, "Einstein recognized rather early signs of dementia praecox [schizophrenia] in his younger son. After many vicissitudes, Eduard was institutionalized in the Burghölzli Hospital in Zürich, where he died in 1965."

many physicists were fascinated by the so-called cathode tubes, large glass cylinders that were fitted with two metal plates connected to the opposite poles of a battery.

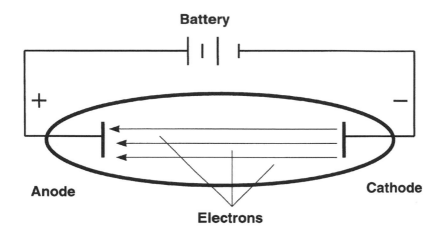

Battery

Anode

Cathode

Electrons

A simplified cathode tube. Electrons move from the cathode to the anode when the tube is connected to a battery.

Röntgen was investigating the properties of electric discharges produced between the two metal plates (the cathode and the anode) when he stumbled onto a very strange effect. Even after covering the tube with black cardboard, he observed that a screen painted with a fluorescent material placed as far away as six feet from the tube would glow in the dark every time a discharge flew across the cathode tube. Something emanating from the cathode tube was irradiating the fluorescent screen.*

Excited by this discovery, Röntgen immersed himself in his laboratory for two weeks, exploring the properties of this unknown kind of radiation, the "X" rays. He showed that the rays were not electrically charged, since magnets did not deflect them. To his amazement, he discovered that the rays could expose a photographic plate wrapped in black paper and could penetrate wood and skin,

---

*A fluorescent material has the property of glowing while being irradiated. This is to be contrasted with phosphorescent materials, which continue to glow even after being irradiated.

exposing metal blocks inside a wooden box and the bones of his wife's hands. His discovery caused an immediate sensation. The whole world was captivated by these amazing rays that had the power to go through materials that reflected light. When a reporter asked Röntgen, "What did you think?" Röntgen answered, "I did not think; I investigated." To the question "What are your rays?" he confessed, "I do not know!"

In 1912, Max von Laue, a pupil of Planck's who in that same year wrote the first textbook on special relativity, proved that X rays were invisible electromagnetic radiation of very high frequency (short wavelength). The radiation was produced when the electrons making up the discharge collided against the anode (the receiving plate) or against the glass wall of the cathode tube.* Shooting X rays at a crystal, Laue showed that the rays behaved like other forms of electromagnetic radiation when forced to pass through small apertures; they diffracted, that is, the overlapping waves interfere after passing through the aperture, creating an alternating pattern of bright (waves reinforce each other) and dark (waves cancel each other) spots on a photographic plate. Soon after Laue's discovery, the English physicist W. H. Bragg (1862–1942) showed that the distances between the bright spots in the diffraction pattern could be used to study the regular geometric structure of the crystal itself. More recently, biologists used X rays to unveil the double helix structure of the DNA molecule. X-ray astronomers now probe the Universe through X-ray emission from distant galaxies and other sources. Research that was initiated with absolutely no intention of being "useful," inspired solely by the curiosity of understanding electromagnetic radiation, became a fundamental tool in medicine as well as in industry. As remarked by Sheldon Glashow, "Today's discovery becomes tomorrow's tool." No fewer than five Nobel prizes have been awarded for research connected with X rays, including, of course, Röntgen's, Laue's, and Bragg's.

A year after Röntgen's discovery, the French physicist Henri Becquerel started an investigation as to whether sunlight could induce certain minerals to become phosphorescent. He thought he

---

*Accelerated charges radiate electromagnetic waves.

could show that phosphorescence had something to do with X rays by placing a phosphorescent mineral on top of a photographic plate wrapped in black paper. The Sun would induce X rays to be emitted by the mineral, which would then expose the photographic plate.* He was quite satisfied when the developed plate showed it had been exposed. When they glow, phosphorescent materials produce not only visible light but also X rays, Becquerel concluded.

A few days later, trying to add dramatic content to his demonstration, Becquerel placed a copper cross between the mineral and the photographic plate. The Sun refused to come out that day, and he stored the mineral, copper cross, and wrapped photographic plate in a dark desk drawer. For some unknown reason, about a week afterward he developed the photographic plate stored in the drawer. He could hardly believe his eyes when he saw the dark image of the cross clearly imprinted on the exposed photographic plate! Sunlight had nothing to do with whatever was emitted by that mineral to expose the photographic plate. So much for his initial conclusions. He then showed that the "rays" emitted by the mineral were not the same as Röntgen's. Since his mineral contained uranium, Becquerel called his rays "uranic rays." Radioactivity was officially discovered!

Two years after Becquerel's discovery, Pierre and Marie Curie showed that several other minerals, including thorium and radium, emitted similar rays. Further work established that there were three different kinds of rays emanating from these elements, which were called alpha, beta, and gamma rays. The next task was to find out what these rays were made of.

During these busy five last years of the nineteenth century, German and French physicists discovered new and surprising phenomena that no one could have guessed existed. Now it was England's turn. In January 1896, Ernest Rutherford, a young New Zealander working with Joseph John Thomson (J.J. to his students) at the famous Cavendish Laboratory in Cambridge, England, wrote to his fiancée:

---

*It turns out that phosphorescence is a chemical process, which has nothing to do with X rays or other forms of radiation.

The Professor [J.J. Thomson] has been very busy lately over the new method of photography discovered by Professor Röntgen. [Thomson] is trying to find out the real cause and nature of the waves, and the great object is to find the theory of matter before anyone else, for nearly every professor in Europe is now on the warpath.

Physics had become very competitive. Thomson was trying to prove that cathode rays were electrically charged. Since electric charges are affected by electric fields, cathode rays had to be too, if they were indeed made of electrically charged particles. Scientists had discarded this possibility because they failed to prove that the rays could be deflected by an electric field.

Here is the mark of the great scientist. Thomson realized that unless the air (or gas) inside the glass tube was efficiently evacuated, no effect could be seen; the gas was acting as a shield against the action of the applied electric field. This is why previous attempts were unsuccessful. Once Thomson succeeded in creating a "good vacuum," he observed that electric fields did deflect cathode rays. By combining the deflections produced by both electric and magnetic fields, Thomson was able to show that cathode rays consisted of negatively charged particles.

He deepened his studies of the particles, or "corpuscles," by showing that they appeared in many different materials with exactly the same properties, such as mass and charge. They were characteristic of all matter. He concluded that atoms were not indivisible, as generally believed then, since they release negatively charged particles when subjected to electric forces. The search for the basic building blocks of matter had to go a step further. He also showed that the particles were at least a thousand times lighter than the lightest atom, hydrogen, isolated for the first time by Henry Cavendish in 1766. On April 29, 1897, Thomson announced his discoveries to the Royal Institution. The first elementary particle, the *electron*, had been discovered.

If the negatively charged electrons were part of atoms, and matter is electrically neutral, then atoms must have a positively charged component as well. Physicists now started to ponder the structure of atoms. Clearly, since the electron was so light, the positive component carried most of the atom's mass. Naturally the first atomic models assumed that the positive charge occupied most of the atom's volume. Hantaro Nagaoka, in Tokyo, proposed a "saturnian model," wherein the negative electrons orbited around a large, spherical, positively charged nucleus, like Saturn with its rings. Thomson proposed a "plum pudding model" (a very British idea), in which the positive charge was distributed throughout the atom while the electrons were like small plums stuck in the pudding. Both models were incorrect. We now know that the positive charge is concentrated in a very small nucleus, which indeed carries practically all of the atom's mass. The electrons orbit far away, making the atom mostly empty space. For example, if we magnified an atomic nucleus to the size of a tennis ball, the electrons would be found about five hundred yards away!

Thomson's former student, Rutherford, made this groundbreaking discovery in 1911. He established the modern picture of the atom after a series of remarkable experiments performed while he was in Manchester, England. Radioactivity was a form of spontaneous transmutation of heavy atoms. When a radioactive atom decays, it is transformed into an atom of a different chemical element. Furthermore, radioactivity is a completely random process. It is impossible in principle to predict, say, when an alpha particle will be emitted by a radioactive nucleus; all we can give is the probability of its happening at some given time. The probabilities used by Boltzmann to describe the collective behavior of many atoms now appeared to be ruling the behavior of the atoms themselves.

In 1911, a young Danish physicist named Niels Bohr came to Manchester to work with Rutherford. On hearing of the new model of the atom, Bohr immediately set to work, trying to understand it in more detail. The more he thought about the problem, the more he realized that classical physics couldn't make head or tail of Rutherford's model. First of all, applying Newton's laws to the electron orbiting the massive but tiny nucleus, like a planet around the Sun, was insufficient to determine the radius of the orbit, i.e., the atom's size. Second, Maxwell's theory stated that a charge in orbital motion would radiate away its energy in higher and higher frequencies as it collapsed onto the nucleus. In other words, classical electromagnetism predicted that atoms were unstable!

Like Einstein before him, Bohr made good use of Planck's quantum hypothesis. He proposed a hybrid model for the atom, combining elements of classical physics with the discreteness inherent in the quantum world. It was a transitional idea, a herald of things to come. As a compromise between a miniature solar system and the discrete nature of radiation, Bohr suggested that the simplest atom, hydrogen, was composed of a positively charged nucleus and a negatively charged electron revolving around it in circular orbits. But—and it was a very big but—not just any orbit was allowed. The electron could be found only at certain distances from the nucleus, or concentric orbits of different radii. The orbit closest to the nucleus, the innermost orbit, was called the ground state of the hydrogen atom. Bohr boldly assumed that the electron couldn't get any closer to the nucleus; for an as yet unknown reason, the quantum nature of small-scale physics guaranteed the stability of the atom, in direct conflict with classical physics.

To this peculiar model of the atom, Bohr added another strange ingredient. He knew that the closer the electron was to the nucleus, the stronger the electric attraction between the two. Thus, an electron in the ground state would need extra energy to move to a higher orbit (an excited state), farther away from the nucleus. Conversely, an electron in a higher orbit would give away energy as it moves closer to the nucleus. Since Bohr knew how to calculate the distances between each orbit and the nucleus, he could also compute the energy of each orbit. He conjectured that in order for an electron to jump to a higher orbit, it had to absorb a photon with exactly the energy difference between the two orbits. The energy of the photon was given by Einstein's formula used in the photoelectric effect ($E = h\nu$). Conversely, an electron jumping to a lower orbit would emit a photon with precisely the same energy as the energy difference between the two orbits. Since photons are electromagnetic radiation, Bohr showed that an excited atom emits electromagnetic radiation as it relaxes to its ground state, while an atom in the ground state absorbs photons as it reaches one of its excited states. Photons and electrons are partners in the quantum waltz.

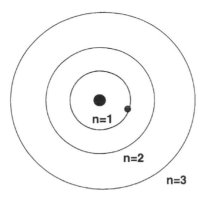

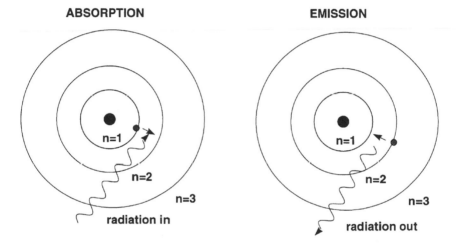

The Bohr model of the atom. Electrons move around the nucleus in discrete circular orbits. Absorption of a photon will induce the electron to jump to a higher orbit. Conversely, a photon is emitted as an excited electron hops down to a lower orbit. In both cases, the energy of the photon is identical to the energy difference between the two orbits.

It was a pretty crazy idea, but it was tremendously successful when checked against experiments. In particular, Bohr could calculate the electromagnetic spectrum of hydrogen, that is, he could predict the frequencies of the emission lines in excellent agreement with the observed spectrum. Finally the mystery of spectra was

solved! Emission lines of specific frequencies are simply photons being emitted by excited atoms as electrons hop down to lower orbits.
Absorption lines (the black lines on spectra) are caused by electrons
"eating up" the photons as they climb up to higher orbits farther
away from the atomic nucleus.

Each orbit was labeled by an integer number n, starting with
$n = 1$ for the ground state. In the world of the very small, the con-
tinuum of classical physics had to be replaced by the discreteness of the
quantum. Integer numbers reentered science, hand in hand with the
physics of the atom. Pythagorean ideas, never completely forgotten,
reemerged with surprising force. In the inspired words of one of the
principal architects of the early quantum theory, Arnold Sommerfield,

> What we are nowadays hearing of the language of spectra is a true
> "music of the spheres" within the atom-chords of integral relation-
> ships, an order and harmony that becomes ever more perfect in
> spite of the manifold variety.

Kepler would have been delighted! The dance of the Universe extends from the very small to the very large.

Despite its initial success, Bohr's atomic theory quickly showed
limitations. It was unable to describe the next element in the atomic
ladder, the helium atom with its two electrons. Eventually, the classical components of his model of the atom, such as the idealization of
the electron and nucleus as small billiard balls arranged in a miniature solar system, had to be abandoned. Yet it was clear that some
part of Bohr's brilliant ideas would have to be included in any future theory, able to embrace more complicated atoms. In the end,
what remained of Bohr's theory was its most revolutionary idea, the
discreteness of the electronic orbits, the emphasis on integers.

MATTER WAVES

In 1921, Einstein (finally!) won the Nobel prize. Even though by
then there was very convincing observational support for his special
and general theories of relativity, his prize was awarded for the explanation of the photoelectric effect using the photon. Surprising as this
may seem to us now, Einstein always said that the photon was his most

revolutionary idea. Millikan's experiments left little doubt that the "heuristic" hypothesis of treating light as a particle worked extremely well. In 1923, a definitive experiment by the American physicist Arthur Compton (1892–1962), proved beyond doubt that X rays bounced off electrons very much like particles would. The so-called *wave-particle duality* of light, sometimes acting as a wave, sometimes a particle, was an inescapable experimental result.* But how could this be? A particle is a small, localized object, whereas a wave is dispersed throughout space: The two are incompatible ways of representing objects with spatial extension. Light behaves as a wave if the experiment is set up to test wavelike properties, such as interference patterns, or it behaves as a particle, if the experiment is set up to test particlelike properties, such as collisions with other particles. Thus, light is neither wave nor particle and yet it is both! It all depends on how *we* choose to probe it.

Two fundamental aspects of the quantum theory arise from this discussion, which are radically different from traditional classical reasoning. First, the images we construct in our minds trying to picture what light is are not appropriate. Language, which represents a verbalization of such mental images, is thus equally limited to deal with quantum reality. As the great German physicist Werner Heisenberg (1901–1976) wrote, "We wish to speak in some way about the structure of atoms, but we cannot speak about atoms in ordinary language." Our language is limited by our bipolar perception of the world, which, of course, includes the wave-particle polarity, something we encountered earlier on in this book, when we discussed how creation myths dealt with the issue of an all-encompassing Absolute. The quantum reality transcends our metaphors.

Second, the observer is not a passive player in the description of natural phenomena. If light behaves as either a particle or a wave depending on how we set up the experiment, then we cannot separate the observer from what is being observed. In other words, in the world of the quantum, the observer plays a crucial role in determining the physical nature of what is being observed. The idea of an objective reality, existing independently of an observer, which was a given in classical physics and even in relativity theory, is lost.

*Light here means any form of electromagnetic radiation, visible or not.

It goes without saying that this new physics disturbed many people. Things got worse in 1924, when a French prince and a relative newcomer to the ever more popular physics scene, Louis de Broglie, suggested in his Ph.D. thesis that the wave-particle duality was not a peculiarity of light, but of all matter. Electrons and protons were also both wave and particle, depending on how we chose to study them through laboratory experiments. Thus, electrons would scatter light as little billiard balls would, but they would also display a striking interference pattern, qualitatively identical to the one displayed by light, if they collided with a crystal. Matter and light could not be pictured in classical terms. In the words of the American physicist Richard Feynman,

> Things on a very small scale behave like nothing you have any direct experience about. They do not behave like waves, they do not behave like particles, they do not behave like clouds, or billiard balls, or weights on springs, or anything that you have ever seen.

Given the bizarre nature of the quantum world, progress could only be made through radically new approaches. In the interval of two years, a brand-new theory of the quantum was invented, the so-called quantum mechanics, which could describe the behavior of atoms and their transitions, without invoking classical pictures such as billiard balls and miniature solar systems.

In 1925, Heisenberg produced his remarkable "matrix mechanics." It didn't include descriptions of particles or orbits, just numbers depicting electronic transitions in atoms. It was a completely new way of describing physical phenomena, a brilliant liberation from the limitations imposed by classically inspired imaging. Unfortunately, it was also notoriously hard to calculate with, even for the simplest atom, hydrogen. However, another brilliant young physicist (there were lots of them around in those days, all in their twenties), the Austrian Wolfgang Pauli (1900–1958), showed that matrix mechanics could be used to obtain the same results as Bohr's model for the hydrogen atom. And then, in 1926, an apparently different way of dealing with atoms appeared, the so-called "wave mechanics" proposed by the Austrian Erwin Schrödinger (1887–1961). Some sense could be made out of the quantum world, but at a cost.

In the spirit of Maxwell's electromagnetism, which described light as waving electric and magnetic fields, Schrödinger was after a wave mechanics that described de Broglie's matter waves. One of the consequences of de Broglie's idea was that if electrons were waves, then it was possible to explain why only certain discrete orbits were allowed. To see why this is true, imagine a string being held by two people, A and B. A jerks it quickly, creating a wave moving toward B. If B did the same, a wave would move toward A. Now, if A and B synchronize their actions, a *standing wave* appears, a pattern that doesn't move left or right and that exhibits a fixed midpoint between them, called a node. If A and B move their hands faster, they will find new standing waves with two nodes, three nodes, and so on. There is a one-to-one correspondence between the energy of the standing wave and the number of nodes.

STANDING WAVES

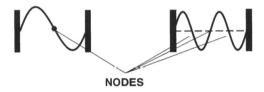

NODES

DE BROGLIE WAVES AROUND NUCLEUS

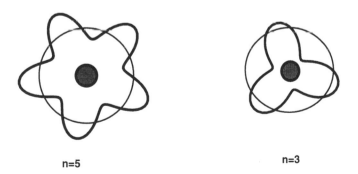

n=5   n=3

Standing waves are characterized by an integer number of nodes. De Broglie pictured the electron as a standing wave around the nucleus. The energy of the electron's orbit has a one-to-one correspondence to the number of nodes.

De Broglie pictured the electron as a standing wave around the nucleus. As such, only certain vibrating patterns would fit in a closed circle—the orbits, each characterized by a given number of nodes. The allowed orbits were identified by the number of nodes of the electron wave, each with its specific energy. Schrödinger's wave mechanics gave a concrete explanation as to why de Broglie's picture of the electron as a standing wave was accurate, but went much further, generalizing this simplistic picture to three spatial dimensions.

In a sequence of six remarkable papers, Schrödinger formulated his new mechanics, applied it successfully to the hydrogen atom, explained how it could be applied to produce approximate answers to more complicated situations, and proved the compatibility of his mechanics with that of Heisenberg's. According to Schrödinger's biographer, W. Moore, it all started during an illicit two-week vacation in the Swiss Alps:

> Erwin wrote to "an old girlfriend in Vienna" to join him in Arosa, while Anny [his wife] remained in Zürich. Efforts to establish the identity of this woman have so far been unsuccessful. . . . Like the dark lady who inspired Shakespeare's sonnets, the lady of Arosa may remain forever mysterious . . . Whoever may have been his inspiration, the increase in Erwin's powers were dramatic, and he began a twelve-month period of sustained creative activity that is without parallel in the history of physics.

Although the last statement is, in my view, a bit exaggerated—Newton's output during the plague years and Einstein's 1905 papers come to mind—it is certainly true that Schrödinger's source of inspiration was quite different from that of Newton's or Einstein's.

The solution to Schrödinger's equation was known as the "wave function," which he thought of as the mathematical expression describing the electron itself. This was in accord with classical notions of how waves evolve in time; given their initial position and velocity, we can use their equation of motion to predict what happens in the future. Schrödinger was particularly proud of the fact that his equation restored some order to the conceptual mess caused by atomic physics. He never liked the idea of the electron "jumping" between discrete orbits. However, it soon became clear that this interpretation for the wave function couldn't be correct. For Heisenberg had just shown that quantum mechanics embraced a new principle, which built yet another wall between the classical and the quantum worlds. This was the famous *uncertainty principle*, which, in its most popular form, states that we cannot know with absolute certainty the position and the velocity (actually, momentum) of an individual particle.

"Wait a minute," you exclaim with indignation. "How can this be? Certainly, if we have more accurate instruments, we can always improve our measurements of a particle's position and velocity. Right?" Wrong! The problem is that the very act of measuring disturbs what is being measured. For example, if we want to see something, we direct some light on it. More detailed imaging needs brighter light of increasingly smaller wavelength (higher frequency); if we try to see a very small object, we must direct light of very small wavelength at it. The problem is that light, like any other wave, carries energy. And, as we know, the smaller the wavelength, the more energy it carries. Thus, in trying to see something with light, we actually add energy to it and kick it out of its original position, giving it some velocity. The better we try to measure its position, the harder we kick it away. Measuring inherently means interfering.

But if we can't specify the location and velocity of objects precisely, then we can't predict their motion precisely. In the world of the very small, the concept of a path becomes fuzzy. That was a terrible blow to Schrödinger. And to Einstein. And to Planck. And even to de Broglie. Schrödinger's frustration exploded in an altercation with Bohr during a visit to Copenhagen:

> *Schrödinger:* If we are still going to have to put up with these damn quantum jumps, I am sorry that I ever had anything to do with quantum theory.

> *Bohr:* But the rest of us are very thankful for it, and your wave mechanics in its mathematical clarity and simplicity is a gigantic progress over the previous forms of quantum mechanics.

The strain caused by these arguments actually made Schrödinger ill. And although Mrs. Bohr showed some compassion toward Schrödinger while he lay ill in bed, Professor Bohr showed no mercy whatsoever, and kept bombarding the weakened Erwin with arguments in support of the reality of quantum jumps.

Well, if the wave function did not describe the motion of the electron, *what* did it describe? Again, physicists were lost. How could the wave-particle duality of matter and light and Heisenberg's uncertainty principle be reconciled with Schrödinger's beautiful (and continuous) wave mechanics? Again, a radical new idea was needed, and again someone had it. This time it was Max Born, who, apart from having the distinction of being one of the principal architects of quantum mechanics, was also Olivia Newton-John's grandfather.

According to Born, Schrödinger's wave mechanics did not describe the evolution of the electron wave *per se*, but rather the *probability* of finding the electron in this or that position in space. By solving Schrödinger's equation, physicists can compute how this probability evolves in time. We cannot predict if the electron will be here or there, but only give probabilities of its being here or there. In quantum mechanics, the probability evolves deterministically, but not the electron itself! The same experiment, repeated under exactly the same conditions many times over, will always give different results. Thus, what quantum mechanics predicts is the probabilities of possible outcomes.

You may be wondering if a probabilistic theory can be of any use in describing nature. The answer is a resounding Yes. Quantum mechanics is tremendously successful in explaining the results of countless experiments probing the very small. In fact, it is the most successful scientific theory ever developed. To its success we owe the invention of a myriad of technological marvels, from transistors and computers to compact discs and digital TV. "Today's discovery becomes tomorrow's tool."

EINSTEIN'S DEMON

Born's interpretation worked like a charm; it enchanted the young and disgusted the old. It demolished the classical notion of a deterministic description of nature. In the world of the very small, the observer plays an active role in determining the physical nature of what is being observed. Laplace's supermind was dead. Furthermore, the results of the measurements can only be given in terms of proba-

bilities. Certainty is replaced by uncertainty, determinism by probabilities, continuous processes by quantum jumps. As you can imagine,
tempers flared, personalities clashed. The Bohr-Schrödinger altercation was followed by many others, the classical clashing with the
quantum. Bohr elaborated his philosophical position in the *complementarity principle*, which states that wave and particle are both equally
possible and mutually exclusive ways by which quantum objects (such
as photons or electrons) will reveal themselves to an observer. They
are complementary forms of existence, only coming into being after
the quantum object comes in contact with an observer. Before contact, the quantum object is neither wave nor particle. In fact, before
contact, we cannot even assert if the object exists. Together with
Heisenberg's uncertainty principle, the complementarity principle
formed the so-called "Copenhagen interpretation of quantum mechanics," largely developed by Bohr during his efforts to elucidate the
conceptual foundation of quantum mechanics.

Whatever its complications, physicists had to choose how to deal
with quantum theory. Did it represent the way things were in the
atomic and subatomic realm, or was it just a temporary theory, waiting for a more profound, and deterministic, formulation? Opinions
varied quite a bit. But soon, the younger generation sided with
Bohr's view that uncertainties and wave-particle dualities did not
just represent our ignorance of a more profound description of nature; they represented how nature is, fundamentally uncertain, fundamentally dual. Paraphrasing the psychologist William James, a
source of inspiration to Bohr, "You can't turn the light on quickly
enough to see how the darkness looks." It is no wonder that when
Bohr was awarded the Danish Order of the Elephant in 1947, he
chose the Taoist symbol for Yin and Yang for his coat of arms, with
the Latin inscription *Contraria sunt complementa*, "opposites comple-
ment each other." Born, Heisenberg, and Pauli sided with Bohr. But
it is perhaps in the writings of J. Robert Oppenheimer, now unfortunately known mostly for his leadership of the Manhattan Project
during World War II, that we find the most lyrical expression of the
universality of complementarity:

> The wealth and variety of physics itself, the greater wealth and vari-
> ety of the natural sciences taken as a whole, the more familiar, yet

still strange and far wider wealth of the life of the human spirit, enriched by complementary, not at once compatible ways, irreducible one to the other, have a greater harmony. They are elements of man's sorrow and his splendor, his frailty and his power, his death, his passing, and his undying deeds.

We know of Schrödinger's dislike for the inherent discreteness of quantum theory. But it was Einstein who became its most vocal opponent. In December 1926, he wrote to Born:

> Quantum mechanics demands serious attention. But an inner voice tells me that this is not the true Jacob. The theory accomplishes a lot, but it does not bring us closer to the secrets of the Old One. In any case, I am convinced that He does not play dice.

To Einstein, the probabilistic description of natural phenomena could not be the final word. There was an objective reality out there, independent of the observer. The inherent "connectedness" between observer and observed typical of quantum theory upset him deeply. He conceded its value as an efficient way to describe the results of experiments dealing with small-scale physics. But there must be a deeper formulation of physics, which could do away with such "incomplete" theory. Quantum mechanics should be incorporated into this more complete theory, but could not serve as a basis for it. Einstein believed that accepting complementarity as an end in itself was accepting defeat in our quest for knowledge.

Einstein tried to find holes in the conceptual foundation of quantum mechanics by posing very astute thought experiments to Bohr and his followers, only to be proven wrong on every occasion.* From 1935 onward, Einstein was pretty much isolated in his opposition to quantum mechanics. But he would not let go. As his biographer Pais remarked, the quantum was Einstein's demon. In his own words,

---

*The search for different formulations of quantum mechanics is far from over. However, in spite of several attempts to incorporate some form of "realism" in a novel formulation of quantum mechanics, experiments still rule in favor of traditional quantum mechanics. We will have to wait and see if Einstein will have the last laugh on this one.

(1931) I am still inclined to the view that physicists will not in the long run content themselves with that sort of indirect description of the real. . . .

(1933) I still believe in the possibility of a model of reality—that is to say, of a theory which represents things themselves and not merely the probability of their occurrence.

(1936) I believe that the [quantum] theory is apt to beguile us into error in our search for a uniform basis for physics, because, in my belief, it is an *incomplete* representation of real things. . . . The incompleteness of the representation leads necessarily to the statistical nature (incompleteness) of the laws.

The Bohr-Einstein debate, which was only interrupted by Einstein's death in 1955, sharpened even more their marked differences. Neither was finally convinced by the other. In my opinion, there was more to it than just discordant views on how to interpret the value of quantum mechanics as a physical theory. Behind the Bohr-Einstein debate were their opposing *beliefs* in what physics was about, and what the physicist's goals were when building theories of nature. Theirs was a "religious war," fed by the two very different (and noncomplementary!) ways by which their scientific creativity was inspired.

To Bohr, the fact that quantum theory worked attested to a fundamental complementarity in nature. To Einstein, the fact that quantum mechanics worked only indicated that it had some element of truth that would be part of a final, more complete theory. There was no reason why one should stop at this plateau. Instead, one should keep searching for a more "uniform basis for physics." Einstein's position was a consequence of the "religiousness" that inspired his science, of his rational mysticism. I will let him speak for himself:

The most beautiful experience we can have is the mysterious. It is the fundamental emotion which stands at the cradle of true art and true science. Whoever does not know it and can no longer wonder, no longer marvel, is as good as dead, and his eyes are dimmed. It was the experience of mystery—even if mixed with fear—that engendered religion. A knowledge of the existence of something we cannot penetrate, our perceptions of the profoundest reason and the most radiant beauty, which only in their most primitive forms are accessible to our minds—it is this knowledge and this emotion that constitute true religiosity; in this sense, and in this alone, I am a deeply religious man.

Einstein called this religious inspiration for science "cosmic religious feeling." He referred to it as the "strongest and noblest motive for scientific research," a fruit of a "deep conviction of the rationality of the Universe," and finding expression in a "rapturous amazement at the harmony of natural law."

These are the words of someone who believed in a deep sense of causation operating in nature, a belief that ran against everything that quantum mechanics stood for. In light of their incompatible positions in relation to what science is, it is no wonder that Einstein and Bohr could never agree, despite their mutual admiration for each other. In any case, their debate serves to prove the point I made earlier, that subjectivity plays a crucial role in the scientific creative process. A scientist's personal beliefs, more often than not, shape the approach and goals of her research: Science carries the mark of its creator. Even in the case where the same science is created independently by two scientists, the presentation and focus of the scientific discourse are always unique. At the root of it all is the "mysterious," to Bohr manifest in the duality and fundamental indeterminism of natural processes, to Einstein in the unity and fundamental order in nature, the "cosmic religious feeling" that so inspired him.

Light carries with it the secrets of relativity and quantum mechanics. It is amusing to contemplate what the simple act of turning the light on entails at the quantum scale. From our limited macroscopic perception, light just appears, immediately inundating the room with its perfectly homogeneous and comfortable glow. In fact, every time we enter a dark room and flip the switch on, the wild quantum waltz of photons and electrons unfolds; turning the switch on means that a current flows through the tungsten filament. The current is made of electrons that collide with the atoms of the filament, causing them to vibrate in countless ways. The energy from the vibrations is dissipated as photons of different frequencies, which appear to us as heat (infrared) and light (visible) coming out of the filament. The things we take for granted!

The worlds of the very fast and the very small challenged and expanded the scientific imagination beyond any reasonable prediction. What would Maxwell and Faraday think of time dilation, length contraction, radioactivity, wave mechanics, and electrons "jumping" between orbits while emitting and absorbing photons? In retrospect, it is remarkable how fast and how much physics changed during the first three decades of this century. True, there were more people working on it, there was more money thrown at it, and technology allowed for an unprecedented improvement in the quality of experimental science. The development of quantum mechanics was painful, imposed on physicists from the outside in, a laboratory-driven revolution. Something had to be done in order to explain the results of all these experiments, which so flagrantly defied explanations based on classical reasoning. It was a revolution built upon many different ideas and by many different personalities, a patchwork of trial-and-error and sometimes desperate measures.

Special relativity was the work of one man, imposed from the inside out, driven by thought alone. Einstein's contributions do not stop here, however. Soon after he concluded his special theory of relativity, he started to think about generalizing it to situations involving accelerated motions. Through another brilliant insight, Einstein realized that acceleration and gravity were intimately related. The outcome of his efforts appeared in final form in 1915 as the general theory of relativity, which deeply revised Newton's other great contribution to classical physics, his theory of gravity. A new era for cosmology was about to start, first from the inside out, and then, through the discoveries of the great American astronomer Edwin Hubble, from the outside in. The Universe was about to become a very large place indeed.

PART V

O

MODELING

THE UNIVERSE

9

INVENTING UNIVERSES

I saw a huge Wheel, which was not in front of my eyes, nor behind, nor beside, but in all places at once. This Wheel was made of water, but also of fire, and was (although I saw its border) infinite.

— JORGE LUIS BORGES

Side by side with the revolution in our understanding of the physics of the very small and of the very fast, the first three decades of the twentieth century witnessed yet another revolution: a new physics of gravity and of the Universe as a whole. In short, a new physics of the very large. Once again, the key intellectual stimulus came from the mind of Albert Einstein. For soon after he completed his seminal work on the special theory of relativity, Einstein started to wonder how it could be generalized to include not only observers moving with constant relative velocities, but also observers moving with varying relative velocities, *i.e.*, accelerated motions.

In an insight that he dubbed "the happiest thought of my life," Einstein realized that there was an intimate connection between accelerated motion and gravity. A "general" theory of relativity, capable of incorporating accelerated motion, implied in a new theory of gravity. As with the thought experiment where he asked how a light wave would appear to an observer moving with the speed of light, Einstein's insight into the inner workings of gravity came from a disarmingly simple image: How would someone who is falling down, say, from the top of a tall building (into a nice deep swimming pool), characterize the gravitational forces around him?

In the same way that special relativity revealed the shortcomings of Newtonian mechanics for velocities approaching the speed of light, Einstein's new theory of gravity revealed the shortcomings of Newtonian gravity for situations involving strong gravitational fields. It was already known that gravity could be thought of in terms of fields, just as with electromagnetism after Faraday and Maxwell. Any mass would have associated with it an attractive field, "a disturbance in space," which would then influence other masses placed close to it. But to say that Einstein simply generalized Newtonian ideas to include stronger gravitational fields is a gross understatement. The new theory of gravity, or general theory of relativity as it is known, provided a radically different conceptual framework to deal with the age-old question as to why objects attract each other.

As opposed to the *absolute* space and time of Newtonian physics, both impervious to the presence of matter, in general relativity spacetime becomes plastic, deformable, responding to the presence of matter in well-determined ways: matter (or, due to special relativity, energy) actually alters the geometry of space and the flow of time. In turn, masses placed in this "bent" spacetime will have motions that deviate from the usual straight motions at constant velocities described by special relativity; they will have accelerated motions. In Einstein's general relativity, the effects of gravity are understood in terms of motions in a bent spacetime.

This intimate relationship between matter and the geometry of spacetime has immense cosmological importance. As Einstein realized soon after he completed his main paper on the general theory late in 1915, if the distribution of matter in the entire Universe could be somehow modeled, then the new theory of gravity should determine the geometry of the Universe as a whole! A new era for cosmology was to begin, the shape of the Universe itself amenable to study through the equations of general relativity. Following Einstein's pioneering efforts, new models of the Universe emerged, mathematical universes based as much on different physical assumptions as on personal prejudice. If you mastered the complex mathematics behind the general theory of relativity, you could play around with building universes on a piece of paper. You could play God on a Tuesday afternoon.

As in other instances in the history of physics past and present, what was missing was data, some indication of what was the correct direction cosmology should take. The whole issue could have remained quite academic if not for another revolution, this time in observational cosmology. In a series of remarkable findings during the 1920s, the American astronomer Edwin Hubble not only settled the age-old question as to whether nebulae were other "island universes" like our own Milky Way (see Chapter 6), but also discovered something even more remarkable: The Universe was expanding. In the space of a decade, the Universe not only grew enormously in size, populated by countless galaxies each with billions of stars, but also became dynamic, with galaxies moving away from each other in all directions in the vastness of cosmic space. Models of the Universe now had to accommodate its inexorable expansion.

Together with Hubble's discoveries, Einstein's new theory of gravity ignited a spark that had lain dormant for quite a long time. With brand-new tools, physicists and astronomers could address, in a rational way, age-old questions concerning the structure and evolution of the Universe as a whole. Cosmology, previously the exclusive realm of theological debate or pseudoscientific speculation, became a true science.

If the Universe is expanding, does it have a center? Was there a beginning? How old is it? How big is it? Will it have an end? Are we to be the helpless victims of a cosmic cataclysm of untold proportions? Can we understand the "Beginning"?

We encountered these questions in the first chapter of this book, when I discussed creation myths. The questions are the same, even though scientists will approach them in ways that are quite different from shamans or priests of different religions. I can't overemphasize the differences between a scientific and a religious approach to cosmological questions. The language is different, the symbols are different, with science ultimately conforming to rigid experimental confrontation as opposed to the relative freedom of religious mythmakers. But the questions are the same, there is no avoiding this obvious fact. And this makes for the unique position cosmology occupies among the physical sciences. No other branch of physics deals with questions of this nature, which can legitimately be asked outside the walls of science.

As for the legitimacy of the answers, well, I speak as a scientist and stick to the rationality of the scientific approach, even though I recognize its limitations. In particular, when it comes to the question of "Creation," our own creativity, scientific or not, encounters a stone wall, and we are forced to recall the old teaching of Plato that "all knowledge is but oblivion." Scientists model the unknown with not much more than logical self-consistency and general physical principles as guidance, while mythmakers try to construct images of that which has no image. The results show a beautiful, albeit limited, universality of human thought when it comes to understanding how the "Absolute" became relative, how the "One" became many.

Scientific models of creation, or cosmogonical models, cannot but repeat some of the themes of creation myths. The world either always has been or appeared at some time in the past, was created from chaos or nothing, or, who knows, is forever created and destroyed in a dance of fire and ice. There are only a finite number of possible answers, which have been visited independently by the scientific and the religious imagination. Perhaps even more important than the answers are the questions, which so clearly reveal what it means to be human, as Milan Kundera expressed in his novel *The Unbearable Lightness of Being*:

> Indeed, the only truly serious questions are the ones that even a child can formulate. Only the most naive of questions are truly serious. They are questions with no answers. A question with no answer is a barrier that cannot be breached. In other words, it is questions with no answers that set the limits of human possibilities, describe the boundaries of human existence.

FREE FALL

In 1907, while still employed by the patent office at Bern, Einstein received an invitation to write a comprehensive review on the special theory of relativity. To make things more interesting, Einstein decided not only to review the growing literature on the topic, but also to add a few new thoughts he'd had on possible extensions to his 1905 papers. As you recall from our discussion in Chapter 7, the special theory was based on two postulates: the principle of relativity, which stated that the laws of physics are the same for observers moving with constant velocities with respect to each other; and the constancy of the speed of light irrespective of the motion of its source or of the observer. For the special theory, the emphasis was on motions with constant velocity. This limitation displeased Einstein, since most motions we deal with in real life involve changing velocities. Clearly, the principle of relativity as used in the special theory was too restrictive; the laws of physics should not be different for observers accelerating with respect to each other. A truly general theory of relativity should encompass all motions, accelerated or not.

As a first step, Einstein started to think of uniformly accelerated motion, that is, motion where the velocity changes at a constant rate. To physicists, the most obvious example of uniformly accelerated motion is that of an object falling due to gravitational attraction, such as an apple falling from an apple tree, or a planet orbiting the Sun. Thus, any extension of the principle of relativity should incorporate gravity. Einstein initially tried to modify Newtonian gravitation to fit special relativity, but was unhappy with the results. And then it happened, his happiest thought in life. In Einstein's own words,

> I was sitting in a chair in the patent office at Bern when all of a sudden a thought occurred to me: "If a person falls freely he will not feel his own weight." I was startled. This simple thought made a deep impression on me. It impelled me toward a theory of gravitation.

To understand the importance of this sudden revelation, we must backtrack a bit. One of the great discoveries of Galileo was his realization that all objects fall at the same rate, independent of their mass. This is why a cannonball and a feather, when dropped from the same height (in the absence of air!), will touch the ground at the same time. Gravity is a very democratic force. Now imagine that an evil scientist (a character in a Hollywood movie, of course) wants to repeat Galileo's experiment, but instead of dropping a cannonball and a feather, he drops you and a cannonball. What will you see as you are falling? Apart from the rapidly approaching ground, you will notice that the cannonball will fall with you, side by side. In fact, if you could not look around (or down), and if there was no air resistance, by only looking at the cannonball you could not tell you were falling; you would not even feel the effect of gravity, your own weight!

You don't believe me? Perhaps a less dramatic experiment will convince you of this. Imagine yourself catching an elevator to come down some fifty floors. As the elevator starts going down very fast you feel yourself lighter, your stomach more and more queasy. The faster the elevator goes down, the lighter you feel. If the elevator simply drops, that is, if it is in free fall you will not feel your weight anymore. You and everyone else in the elevator will be free falling, floating around trying to avoid banging into each other.*

What this image told Einstein was that gravity could be "cancelled" if a proper reference frame was chosen. For example, inside the free-falling elevator there is no gravity, and thus no acceleration; objects moving at constant velocity inside the elevator will remain at constant velocity during the free fall. If they are at rest with respect to each other, they will remain at rest. In other words, inside the free-falling elevator, the principles of special relativity apply. You can see that if objects fell at different rates in a gravitational field this conclusion would be wrong. "Free fall for all" is only possible because gravity is a universal force.

The image also told Einstein something equally important—that for an observer inside a box, without contact with the outside world, it would be impossible to distinguish between the acceleration caused by gravity and an acceleration caused by another force. It is easy to see why this is so, although it does take courage. Assume that you have been locked inside a box that has been launched into outer space without your knowing it. (If you are claustrophobic, take a deep breath and move on. But don't give up; it's only a thought experiment!) The box is attached to a spaceship that is accelerating at a rate exactly equal to the acceleration of gravity at the surface of the Earth. When you come to your senses, a voice coming from a loudspeaker explains that you are now the proud participant in an important scientific experiment. You threaten to sue, but "the

---

*This is exactly what happens to astronauts at "zero G," even though zero G is a misnomer for astronauts in orbit; a spaceship orbiting the Earth is not at zero gravity, but is free falling. Only if the spaceship is very far from any massive body will it really be at (almost) zero G. The lesson here is clear: free fall can simulate zero G, although it isn't the real thing.

GRAVITY REMOVED BY FREE FALL

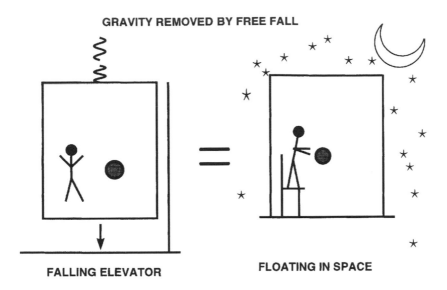

FALLING ELEVATOR **FLOATING IN SPACE**

An observer in free fall doesn't feel the acceleration caused by gravity.

Voice" in the loudspeaker explains that, while half asleep, you happily signed a contract agreeing to take part in this experiment. Having nothing to lose, you begrudgingly agree to go on. The Voice commands you to open a cabinet and get two balls, one made of wood, the other made of lead, and drop them down from a height of three feet. Annoyed at the stupidity of the request, you ask why this is of any importance. "Patience is well rewarded," says the Voice. As you drop the two balls, you notice that they fall together as they should, and record the time it took them to hit the ground in a notebook. (You have been supplied with high-tech equipment to do such measurements.)

The Voice then asks you the question, "Can you use your measurements and tell me where you are?" Remembering your high school physics class, you know how to compute the acceleration from your data. You get the same result you find on Earth and proudly announce to the Voice: "Clearly, since the measured acceleration is identical to that on Earth, I must be on Earth." "Ha, Ha, Ha," laughter echoes inside the box. "You fool! See that little button under the closet? Push it!" As you push the button, the walls retract, revealing a further set of walls to the box, all transparent. You see the spaceship above you and stars, thousands of stars, all around.

And nothing else. You feel a deep loneliness inside, a longing for familiar grounds. "Hey, take me home!" you demand, your voice equally shaken by terror and by the beauty of it all. "Yes, you will go home. But first, you must explain what is going on," says the Voice.

Having no other choice, you start to think. You figure, quite quickly, that the reason why you thought you were on Earth is because the spaceship is accelerating upward, creating a "mock" gravitational force.* Think of an elevator going up, the reverse of free fall; the elevator's acceleration upward feels like extra weight on your feet. This is Newton's third law of motion, the law of equal action and reaction. The elevator's floor pushes your feet up, and your feet push down on the elevator's floor.

You conclude that, for practical purposes, it is impossible to distinguish between an acceleration upward and a gravitational field downward. This is what is known as the *equivalence principle*. Any gravitational field can be mocked by an appropriately chosen accelerated frame. (In this case, the accelerated frame is the box and the spaceship.) You can now understand why Einstein was so excited by this thought: A general theory of relativity that includes accelerated motions is necessarily a theory of the gravitational field. Furthermore, you can always go back to the limit of special relativity (no accelerations) by choosing an appropriate free-falling reference frame. Being a physics lover, you hasten to forgive the Voice, thanking it profusely for allowing you to learn so much about gravity and, as a bonus, to admire the beauty of outer space firsthand. Little do you know there is more to come.**

It takes you quite some time to recover from the shock of your scientific experiment. Although you know nobody will believe you, you invite some friends over to tell them your adventures. While you are telling them the part when the walls retracted and you found

---

*You also realize that in fact the balls don't fall. It is the box's floor that accelerates up to meet them, fooling you into thinking they were falling.

**For those of you who are physics buffs: The equivalence principle is an expression of the fact that the "inertial" mass (m_i), that is, the mass that responds to a force according to $F = m_i a$, is identical to the "gravitational mass" (m_g), the mass that is attracted to another mass according to Newton's law of gravitation, $F = GMm_g/R^2$, where G is a constant known as Newton's constant, M is the mass attracting m_g and R is the distance between the two masses.

yourself alone in outer space, the phone rings. To your surprise and delight, it is the Voice again. Another experiment is being planned, and the Voice needs volunteers. You and a couple of other brave souls eagerly agree to participate.

You and your friends are to be launched into outer space to study the properties of light under acceleration. The plan is to have you alone in one box and your friends in another box. As before, both boxes are transparent and pulled side by side by spaceships. The main difference is that during the experiments your box is to be pulled with constant acceleration, while the box with your friends inside is to move at constant velocity. In other words, during the experiments, you are accelerating with respect to your friends. While you perform the experiments, your friends have to observe what is going on from the point of view of an inertial reference frame.

UNIFORM ACCELERATION SIMULATES GRAVITY

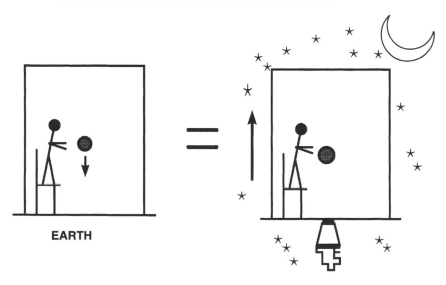

The Equivalence Principle: A uniformly accelerating spaceship can mock the acceleration caused by gravity at the Earth's surface.

The first experiment is simple. The two spaceships start together, moving side by side with the same constant speed. You are asked to shoot a ball with constant horizontal velocity and observe its path, comparing your observations with those of your friends. As soon as you shoot the ball, your spaceship starts to accelerate upward. Thus, although you and the box are given an upward acceleration, the ball, which is not in touch with you or the box, does not feel it. While your friends see the ball moving in a straight line, you see it falling in a curved path like a projectile on Earth, before it hits the opposite wall at some point. By increasing the ball's horizontal velocity, the deflection from the horizontal gets smaller, but is still there. You are not too surprised, since from your previous experience you know that an accelerated frame is just like a mock gravitational field.

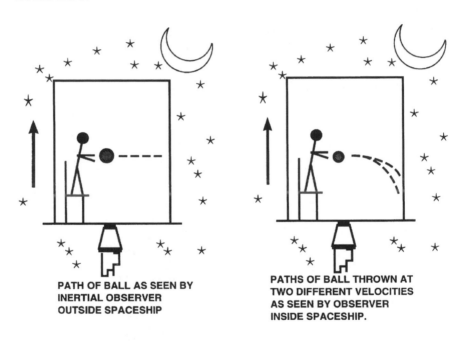

PATH OF BALL AS SEEN BY INERTIAL OBSERVER OUTSIDE SPACESHIP

PATHS OF BALL THROWN AT TWO DIFFERENT VELOCITIES AS SEEN BY OBSERVER INSIDE SPACESHIP.

Path of ball as seen by observers in the outside (left) and in the inside (right) of the accelerating spaceship. The larger the velocity, the straighter the path. Even for light the path would be curved by acceleration, or by the equivalence principle, by gravity.

For the second part of the experiment, instead of firing a ball, you are to shoot red laser light, always horizontally with respect to the floor. For this experiment the spaceship is to impart a pretty large acceleration to your box, so as to mock a very strong gravitational field. Of course, thanks to some as yet unknown technology, you are safely protected from the nasty effects produced by this acceleration. To make things more exciting, the Voice fills your box with a dense fog, so that you can actually see the path of the light beam. Just as with the ball, your friends see the laser light move in a perfectly straight line. And just as with the ball, you see the light beam curve down! You can hardly believe your eyes. The conclusion from this experiment is incredible; since an accelerated frame mocks a gravitational field, light can be bent by gravity! Again, you can see why Einstein was so excited by his thought. This effect was quite unexpected, a direct consequence of the equivalence principle.

Remarkably, Einstein was not the first person to propose that light could be bent by gravity. For Newton, since he believed light was made of tiny corpuscles, it too could be deflected by gravitational attraction. As he wrote in his treatise on light, *Opticks*, "Do not bodies act upon Light at a distance and by their action bend its Rays; and is not this action strongest at the least distance?" The eighteenth-century French astronomer Pierre Laplace, following a similar hunch, conjectured that for heavy enough stars, the gravitational pull would be so strong that not even light would be able to escape, an idea that later, with the advent of general relativity, germinated into the modern theory of black holes. But Einstein took these ideas to the extreme, as he painstakingly developed a new theory of gravity.

In the 1907 review paper, Einstein announced the equivalence principle and some of its physical consequences. Apart from the bending of light by gravitational fields, Einstein derived another effect involving light, the so-called *gravitational redshift*. He reasoned that in a strong gravitational field the sources of electromagnetic radiation, *i.e.*, vibrating electric charges in some material, would have their wavelengths affected; the stronger the field, the larger the wavelength, as if the field were stretching the electromagnetic waves being produced. Since red has the largest wavelength of the visible spectrum, the effect became known as gravitational redshift. Light emitted in a strong gravitational field is shifted toward the red end of the visible spectrum. As Einstein wrote in the 1907 review, "It follows . . . that light coming from the solar surface . . . has a longer wavelength than the light generated terrestrially from the same material on Earth."

Another way of thinking about this effect is to imagine that in the presence of a strong gravitational field atoms vibrate more slowly (smaller frequency), thus producing electromagnetic waves of longer wavelength. Since the vibrational frequencies of atoms are extremely regular, we can think of them as little clocks, ticking at furious rates. Thus, the gravitational redshift is equivalent to a slowdown of clocks; gravitational fields affect the flow of time, the stronger the field, the slower the flow! If we go back to the light clock introduced in Chapter 7, the effect of a strong gravitational field would be to increase the time interval between a "tick" and a "tock." In a rough sense, it will be harder for the light pulse to bounce up and down between the mirrors in the presence of a gravitational field, just as it is harder for you to jump up in a strong gravitational field.

In marked contrast to his 1905 papers, all impeccably developed, Einstein's 1907 results concerning gravity relied on somewhat clumsy approximations, some of which produced incorrect answers, even though the overall results were qualitatively correct. Einstein knew he had a major challenge ahead. And indeed, it occupied him on and off for the next eight years, until he arrived at the general theory as we know it, late in 1915. It was a long and tortuous road, with many false starts and wrong turns. But Einstein's complete faith in his ideas remained absolutely unabated during this whole time. He knew his intuition had to be right, the only problem being how to formulate it in a mathematically consistent way. Physicists who rely on their intuition to do their work can relate to this situation, when your ideas are way ahead of your mathematics, much to your frustration. You know where you want to go, or at least have a good idea of the right direction to take, but feel completely stuck because it is often very hard to translate ideas into equations. And there is no way out; before you can do this, nobody will truly believe you. Ideas are much harder to understand than mathematics. But Einstein's efforts paid off. As we'll see, the general theory of relativity is one of the greatest achievements of the human intellect.

CURVED SPACES

From December 1907 to June 1911, Einstein didn't write a word on gravitation. The reason for his silence lies in his struggle to "under-

stand" quantum theory and the dual nature of light, his constant
"demons." Partially defeated, in 1911 Einstein returned to the equivalence principle he had formulated in 1907. He realized that the bending of light by a gravitational field could actually be observed if the light
from a distant star passed close enough to the Sun. During a solar
eclipse, with the sunlight temporarily blocked, astronomers could measure the position of the star and compare it with measurements taken
when the Sun was not between the star and the Earth. If there was a deflection of the star's light by gravity, the conclusion would be clear;
light is indeed bent by gravity. But confirmation didn't come easy.

An expedition to Brazil to observe a solar eclipse was organized
in 1912 by the British astronomer Charles Davidson, but bad
weather precluded any viewing. Then, in 1914, a German expedition sponsored by the industrialist and weapons manufacturer Gustav Krupp set off to Crimea to observe the eclipse of August 21.*
However, Germany declared war on Russia just a few weeks before
observations were to be made, and the Russian authorities confis-
cated the astronomical equipment and temporarily arrested some
of the astronomers. The question of the bending of light by the Sun
had to wait until World War I was over.

Between 1911 and 1915, Einstein struggled to formulate general
relativity in a mathematically consistent way. The problem was that
the new theory of gravity asked for a completely novel way of interpreting the interplay between matter and the geometry of spacetime.
We can understand this by going back to our experiment with the
bending of light. When inside the box you noticed, to your astonish-
ment, that gravity bent the path followed by the light rays. Einstein realized that there is another way of interpreting this phenomenon,
which is the cornerstone of the new theory of gravity. Instead of saying that gravity bent the path of light, we can equally say that light followed a curved path because space itself was bent! The curved path
was the shortest path possible in this distorted geometry, and, as the
French mathematician Pierre de Fermat showed in the seventeenth
century, light always travels the shortest path between two points.

---

*Gustav Krupp belonged to the same Krupp family largely responsible for the rear-
mament of Germany during Hitler's government. It is a tragic irony that the same
family that sponsored an expedition trying to vindicate the theories of a Jewish scientist was later involved in exploiting Jewish workers under subhuman conditions.

Let us think about this for a moment. When we say that light always travels the shortest possible path between two points, *i.e.*, a straight line, we are basing our observations both on Fermat's principle and on what is called flat Euclidean geometry, the one we know and love from high school. Flat geometry is best understood by referring to the top of a flat table. As we all know, the shortest distance between two points on the tabletop is a straight line. If we shoot laser light parallel to the table's surface, its path will be very very straight. We can also play with triangles, squares and circles, all drawn on the flat tabletop. The results of different manipulations involving these figures and lines is what we call Euclidean geometry, organized (but not entirely created) by Euclid around 300 B.C. Euclidean geometry is formulated on this flat space and encompasses many famous results, such as the sum of the internal angles of any triangle is 180 degrees, or that one and only one parallel line can be drawn passing by a point exterior to another line.

Flat Euclidean space does not have to be two-dimensional like the tabletop. It can have any number of dimensions, although we have trouble seeing more than two. We can see the flatness of the tabletop because we can look at it from the "outside," that is, from a three-dimensional perspective. To see a three-dimensional flat space, we would have to exist and function in four spatial dimensions. But what the eyes can't see the mind can grasp, and it is quite easy to study flat geometry in any number of dimensions through the use of mathematics.

What happens if the tabletop is not flat? Well, the first thing that comes to mind is that the shortest distance between two points will not be a straight line anymore. Think of a large rubber sheet that has been carefully stretched so as to be perfectly flat and square. Now place a heavy metal ball in the center of the rubber sheet. The ball's weight will deform the rubber sheet much in the same way a heavy mass deforms the geometry of space, although we must keep in mind that the rubber sheet is a two-dimensional curved surface, as opposed to the real three-dimensional space we are interested in. Still, the analogy holds very nicely. (We will forget about altering the flow of time for now.)

If we now throw marbles on the deformed sheet, they will move on curved paths. Close enough to the large mass, the marbles will follow circular or elliptical orbits before friction causes them to spiral into the cavity. If we know the geometry of the deformed rubber sheet, we can write equations describing their curved motions. Neglecting friction, and extrapolating to three dimensions, these are the motions of small masses in the presence of a large mass, for example, planets or comets orbiting the Sun.* Einstein's general theory of relativity substitutes Newtonian "action at a distance" by motion in curved space. What we perceive as accelerated motion caused by gravity is simply motion in curved space. If we know the geometry of spacetime, we can predict the motions of objects or light in it. Conversely, the presence of massive objects deforms spacetime from the flat spacetime of special relativity. Paraphrasing the American physicist John Archibald Wheeler, "matter tells spacetime how to bend, and spacetime tells matter how to move."

As he worked on his theories, Einstein asked his old classmate and friend Marcel Grossman for help with the math. Curved geometry wasn't exactly a popular topic, the kind we learn from textbooks. (I will let you decide if textbooks are ever popular or not.) And if Einstein couldn't understand curved geometry, he couldn't formulate general relativity in mathematical form. After the approximations of the 1907 and 1911 papers, it was time to get more precise. Luckily enough, during the nineteenth century, a few brave mathematicians had tackled the problem of curved geometry in detail. They realized that the results of Euclidean geometry break down as soon as space is bent. Furthermore, they showed that the simplest *non-Euclidean* geometries came in two types. There could be spaces of positive curvature, like the two-dimensional surface of a ball, or spaces of negative curvature, like the two-dimensional surface of a saddle. More complicated geometries could be broken down into combinations of these two types. Clearly, in both curved spaces the shortest distance between two points is not a straight line. The departures from Euclidean geometry go opposite ways in the two curved geometries. For example, while the sum of internal angles of a triangle is larger than 180 degrees in a positively curved space, it is smaller than 180 degrees in a negatively curved space. To see this, take a common globe and trace a triangle connecting two points at the equator to the north pole. The

---

*Or apples falling down. Close enough to the large mass, the paths are vertical lines and the motions are toward the center of the large attracting mass.

sum of all three angles will be larger than 180 degrees. In fact, the two angles at the equator alone make 180 degrees.

Flat and negatively curved spaces are called *open*; if you walk in the same direction, you will never come back to your starting point. Positively curved spaces are called *closed*; if you walk in the same direction, you will eventually come back to your starting point, as you can easily visualize on the two-dimensional surface of a globe. It follows that closed geometries are finite; they have a finite volume. And yet, they are unbounded, they have no boundaries. This is a somewhat strange concept, because we are used to thinking of finite spaces as being precisely spaces with a boundary, like states in the political map of a country. How could a finite space have no boundaries?

The first thing to remember is that a circle (a one-dimensional finite space) has no beginning or end. A circle has no boundaries, and yet it is finite. Now imagine the surface of a sphere. If we couldn't look at it from the "outside" and were constrained to live on it as two-dimensional beings, we could crawl and crawl along the surface and would never find its boundary. A closed geometry is finite with no boundaries.

After understanding non-Euclidean geometry, Einstein still had to accomplish the gigantic feat of incorporating it into physics in such a way that the final theory would be consistent with the principle of equivalence (in free fall gravity is cancelled which means that special relativity is valid nearby) and with the most sacred law of physics, conservation of energy and momentum. After many frustrated attempts, in the fall of 1915 Einstein obtained the equations of general relativity in final form.

There are two equations, one relating the curvature of spacetime to the presence of mass ("Einstein's equation"), and the other describing motions in a curved geometry ("Geodesic equation"). Applying them to the problematic orbit of Mercury, he obtained a spectacular agreement with observations. The orbit of Mercury precessed because of its proximity to the Sun, where geometry was mostly curved.

Of the two other immediate predictions from his theory, gravitational redshift and bending of light, the latter was the only one readily accessible for observations. Einstein's new calculation for the angle by which light from a star would deviate from a straight path differed from his 1907 prediction by a factor of two.

TWO DIMENSIONAL GEOMETRIES

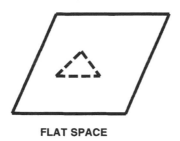

FLAT SPACE

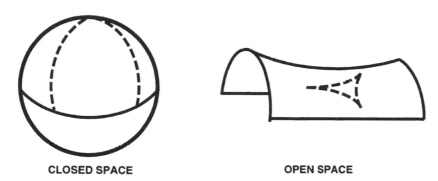

CLOSED SPACE **OPEN SPACE**

Two-dimensional non-Euclidean geometries: (top) Flat geometry with triangle. The sum of the angles of a triangle is 180 degrees. (bottom left) Closed geometry with triangle. The sum of internal angles is larger than 180 degrees. (bottom right) Open geometry with triangle. The sum of internal angles is smaller than 180 degrees.

In 1919, World War I finally over, the British astrophysicist Arthur Eddington organized two expeditions, one to Sobral in northeastern Brazil and the other to Principe Island off the coast of Equatorial Guinea in West Africa to observe the light of a star during a solar eclipse. The results, although initially not crystal clear, were convincing enough; light is bent by gravity in close agreement with the prediction of Einstein's new theory.

The confirmation of Einstein's general theory turned him into an instant international celebrity, the most famous living scientist.

The physicists that understood the theory, initially only a handful, were taken aback by its tremendous beauty and power. The physicists that didn't, or couldn't, understand it, deemed it the fancy of a crazed (or sometimes just "Jewish") mind. The press went absolutely overboard with stories of bent space and time. The public responded accordingly, mystified by this new theory of gravity that shook the very foundations of the beloved Newtonian view of the world. Einstein became a God-like figure, the man who understood the fabric of the Universe like no one else before him. The scientist became the prophet.

Desktop Universes

Soon after Einstein completed the 1915 paper on general relativity, he started to wonder about the implications of his new theory for our understanding of the Universe as a whole. He reasoned that since the equations of general relativity describe the bending of spacetime due to the presence of matter, if the total distribution of matter in the Universe were somehow known, the same equations could be solved to provide the geometry, or shape, of the Universe as a whole. I don't think I have to stress what a bold, and fascinating, move this was. But I'll do it anyway. Up to then, general relativity was confined, if not in theory at least in practice, to effects somewhat local to our own solar neighborhood, such as the anomalous orbit of Mercury or the bending of starlight by the Sun. But now Einstein was suddenly extending the domain of general relativity to the whole Universe! Gravity, the ultimate cosmic glue, became also the Cosmic Artisan.

Einstein was the first to recognize the potential for controversy his ideas carried. In a letter to his friend and physicist Paul Ehrenfest, a few days before presenting his new cosmological model to the Prussian academy early in 1917, Einstein wrote: "I have . . . again perpetrated something about gravitation theory which somewhat exposes me to the danger of being confined in a madhouse." Einstein's proposal inaugurated a new era in cosmology, the application of general relativity to the Universe as a whole in order to study its structure and evolution. Once again scientists would wonder about the shape, size, age, and future of the Universe, questions that had remained somewhat on the back burner for over a century. Whenever there is new physics, there is new cosmology.

One of the first issues to be addressed was whether the Universe was bounded and thus finite, or infinitely vast. As we have seen it, it was a slow and painful transition from the closed, Earth-centered Aristotelian Universe that dominated medieval theology, to the closed, Sun-centered Copernican (and Keplerian, and Galilean!) Universe of the Renaissance. Cries for an infinite Universe populated by an infinite number of worlds like our own, as uttered by Giordano Bruno, were either silenced by the Church or fell on deaf ears. Newton radically changed this situation through his proposal of an infinite and thus open Universe, where universal gravitation was kept in balance by an ever-active God. But an infinite Universe, populated by an infinite number of stars, presented some additional very serious problems. As Halley argued during a Royal Society meeting in 1721, and Kepler a century before him, an infinite Universe with an infinite number of stars distributed at random would be ablaze with light, day or night alike. The Newtonian solution, invoking God's interference, was no longer a popular solution, as Deism confined the role of God to Creation alone, the Cosmic Watchmaker.

So why, in an infinite Universe, is the night sky dark? This paradox, reformulated in 1823 by the German physician Heinrich Olbers, became known as *Olbers's paradox*. At the time, it was generally believed that the solution was interstellar absorption; gas clouds permeating the Universe absorbed the light from distant stars, filtering the amount of light finally reaching us. Unfortunately, as pointed out by William Herschel's son John in 1848, absorption was not the answer, as gas clouds that absorbed light would also remit it, re-creating the problem. It was a remarkable embarrassment, that scientists couldn't understand one of the most obvious of everyday phenomena, the darkness of the night sky. The final solution to Olbers's paradox had to wait for the discovery that the Universe had a beginning and thus a finite age. But before this could happen, Einstein had to open the way with his own attempt at a solution to the problem: a static, and curved, Universe.

Einstein, like most people in 1917, saw no reason to postulate a Universe that changed in time. To be sure, there were small-scale

motions, local displacements of stars, but these didn't show any overall trend. There was no compelling evidence that large velocity
motions existed in the Universe, although in 1912 the American
astronomer Vesto Slipher measured the radial velocity of a spiral
nebula, that is, the component of the nebula's velocity in our direction. Using the Doppler effect, to be discussed shortly, Slipher found
that Andromeda was approaching the Sun at about 300 kilometers
per second, a very large velocity. By 1917, Slipher had measured the
radial motions of other nebulae and found that most were actually
not approaching us, but receding from us. But European theoretical
physicists, Einstein included, had little or no contact with the rapidly
growing American astronomical community. And even among American astronomers, Slipher's measurements were not without controversy. A static Universe was still a very compelling hypothesis.

The Universe was not only believed to be static, but also with
most matter roughly centered around the Milky Way. All objects observed in the night sky, stars and nebulae alike, were thought to be
part of the Milky Way, which stretched outward into the empty vastness of infinite space. The debate over whether some nebulae were
other "island universes" was still open, but the arguments were
mostly not in favor of this hypothesis. In a few years, the Universe
would become a very different place.

Einstein was not comfortable with the notion of an infinite Universe with a finite amount of matter. He believed that a spatially
bounded, and thus finite, Universe was a much more natural choice
from the point of view of general relativity. The geometry of a finite
Universe is uniquely determined by its total mass, he thought. And
what Universe would be simpler than a Universe in which matter is
distributed equally on average. He thus formulated what is known as
the *cosmological principle*, which states that the Universe on average
looks the same everywhere in all directions. That is, that the Universe is homogeneous (same everywhere) and isotropic (same in all
directions); there is no preferred point in the Universe.

Once homogeneity and isotropy are assumed, solving Einstein's
equations becomes a much simpler task; the geometry is determined by only one parameter, which gives a measure of its curvature, its *curvature radius*. *"On average"* here is very important. Of
course Einstein recognized that the Universe was lumpier in some

places than others. But on average, for large enough volumes, the Universe looks the same; and because his is a static universe, the distribution of matter doesn't change in time. In Einstein's finite universe, the total density of matter, that is the total amount of matter over the total volume, is constant. And hence, so is the geometry.

Armed with these hypotheses, Einstein obtained his cosmological solution. It described a static, finite universe of closed geometry, a three-dimensional generalization of the surface of a sphere.* As such it had a radius, which, as expected, was determined by the total mass of the Universe. As he proudly announced in 1922, "The complete dependence of the geometrical upon the physical properties becomes clearly apparent by means of this equation." But this solution came with a high price tag. In a finite and static Universe, and with gravity being an attractive force, matter tends to collapse upon itself. A finite and static Universe, filled with a constant density of matter, simply could not be.

In order to keep his universe static, Einstein included, by hand, an extra term into the equations of general relativity, which he initially dubbed a "negative pressure." It quickly became known as the *cosmological constant.* This term was allowed mathematically, but had absolutely no fundamental physical justification, although Einstein and others tried very hard to find one. It clearly detracted from the formal beauty and simplicity of Einstein's original equations of 1915, which achieved so much without any arbitrary extra constants or assumptions. Basically, the cosmological constant amounted to a cosmic repulsion term that was chosen to exactly balance the tendency of matter to collapse upon itself. Einstein knew that its sole reason to exist was to produce a static and stable finite Universe. It didn't occur to him that a changing, dynamical Universe was also a possibility, quietly hiding behind his equations. Even Einstein could sometimes miss the boat.

At about the same time that Einstein proposed his cosmological model, another model appeared in the literature, proposed by the

---

*It has been asserted by the American Medical Association that trying to "see" three-dimensional closed geometries can raise your blood pressure to dangerous levels. They recommend sticking to two-dimensional surfaces, such as the surface of a balloon.

Dutch physicist and astronomer Willem de Sitter (1871–1934). De Sitter had been deeply impressed with the ideas of general relativity since he first read Einstein's 1911 paper, and he immediately started to study the theory, with a focus on how to bring astronomical evidence into it. De Sitter's cosmological solution was quite bizarre. He showed that apart from Einstein's static solution with matter and a cosmological constant, it was possible to find a solution with no matter and a cosmological constant. A universe with no matter in it was clearly an approximation to the real thing, as de Sitter knew very well.

It turned out that de Sitter's model had a very curious property. In his matterless universe, any two points in it moved away from each other with a velocity proportional to the distance between them. Points at a distance $2d$ moved away from each other twice as fast as points at a distance d. De Sitter's universe had motion, even though it was empty! The cosmic repulsion fueled by the cosmological constant stretched this universe apart. Thus, while Einstein's universe had matter without motion, de Sitter's universe had motion without matter; the models were, in a certain sense, complementary to each other.

Since de Sitter's universe was empty, its expansion could not be perceived by any observer. But in the early twenties, mostly through the work of de Sitter, the British astrophysicist Arthur Eddington, and others, some physical properties of this curious empty universe were discovered. First, if a few grains of dust were sprinkled into de Sitter's universe, they would, like the geometry itself, be scattered away from each other with velocities that increased linearly with distance; they would be dragged along by the geometry, somewhat like corks floating down a river. Another often-used analogy is that of a raisin bread being baked. As the bread grows, all raisins move away from each other.* Granted, grains of dust are matter, and de Sitter's original solution was for an empty Universe. Still, a few little grains scattered here and there in the vast empti-

*Corks floating on a river or runaway raisins in baking breads don't quite do justice to the expansion of the Universe. But sometimes an approximate image is better than no image at all. There is a fine line between a useful image and a confusing one, as my colleagues who are brave enough to venture into scientific popularization know only too well.

ness of space can't possibly spoil de Sitter's assumptions. This is the power of modeling in action. But if the velocities increased with distance, there would be grains so far from each other that they would be receding with velocities approaching the speed of light! And since no information can travel faster than light, any point farther away would be invisible. Thus, each grain would have a *horizon* surrounding it, a boundary beyond which the rest of the Universe is invisible. As Eddington put it, "the region beyond . . . is altogether shut off from us by this barrier of time." This inherent limitation to how much we can know of the Universe disturbed a lot of people during the early twenties. But since de Sitter's model was only a crude approximation . . .

Another consequence of de Sitter's model is even more fascinating than the existence of horizons. If instead of grains of dust we placed a few light sources, such as stars, in de Sitter's universe, they would, like the grains, move away from each other with speeds proportional to their distances. It had been known then for quite some time that waves are affected by the motion of their sources. We are all familiar with this effect from our experiences with sirens and horns; the sound wave of an ambulance's siren, when approaching you, has a higher pitch (shorter wavelength) than if the ambulance is not moving with respect to you. If the ambulance is moving away from you, the pitch is lower (longer wavelength). This is the famous *Doppler effect*, named after the Austrian physicist Johann Christian Doppler, who in 1842 proposed that it was due to a shift in the wavelength of the wave being emitted by a moving source.

Doppler's ideas were confirmed in a most dramatic way by the Dutch meteorologist Christopher Buys-Ballot in 1845. Through his connections with the government, Buys-Ballot borrowed a steam locomotive and a stretch of railroad for a couple of days. His idea was simple. Get a few musicians from the brass section of an orchestra to stand on a flatcar and have them play the same note as the train passes by a group of music experts who could distinguish notes by ear with great accuracy. Asking the experts for the change in pitch of the note as the train cruised by at different speeds, Buys-Ballot could test Doppler's formula. The poor musicians had to blast their instruments against the hellish noise and smoke coming from the steam locomotive, but after a few frustrating trials, the experts finally confirmed the change in pitch as predicted by Doppler's formula.

STATIONARY SOURCE

APPROACHING SOURCE

RECEDING SOURCE

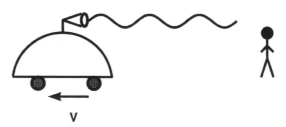

Doppler Effect: For a source moving toward an observer the wavelength decreases (middle), while for a source moving away from an observer the wavelength increases (bottom).

The Doppler effect always makes me think of the rare occasions when I heard my father playing the accordion, an old Scandalli that had been in the family for who knows how long. He would sit at the edge of the bed and vigorously move the two ends in and out, literally squeezing the music out of the instrument, the bellows revealing its bright red color as it expanded and contracted in a rhythmic dance. Expanding, the sounds would shift to lower tones, contracting, the sounds would shift to higher tones. His fingers rapidly cruising over the keyboard, creating music at once new and old, full of magic. Unlocked by the expansion and contraction of the bellows, the secrets of the Universe revealed for a fleeting moment, the dancing Shiva and the harmony of the spheres, the gleam in my father's smiling eyes.

Just like sound waves, light waves are also affected by the motion of their source. A source receding from you will have its wavelength increased, while a source approaching you will have its wavelength decreased, like the bellows of my father's accordion. Since red has longer wavelength than blue, receding light sources will have their spectra shifted toward the red, while approaching sources will have their spectra shifted toward the blue. In de Sitter's universe, all light sources will be moving away from each other; if you sit on any one of them, all other light sources will be redshifted. Furthermore, the amount of redshift can tell us what is the speed of the receding light source. The larger the shift to the red of the spectral lines, the faster the source is receding from us. De Sitter's universe offered the first indication that, in an expanding universe, spectra of distant light sources are redshifted. But since de Sitter's universe was only a crude approximation . . .

While these properties of de Sitter's solution were being found, in Saint Petersburg, Russia, Alexander Alexsandrovich Friedmann, a meteorologist-turned-cosmologist, chose to follow a completely different road. An excellent mathematician, Friedmann quickly learned the technicalities involved in the formulation of general relativity. Inspired by Einstein's cosmological speculations, Friedmann started a program to find other possible cosmologies, perhaps less restrictive than Einstein's. Clearly, he reasoned, Einstein included the cosmological constant in order to keep his model of the Universe static. But why must it be so? Perhaps inspired by the ever-changing weather, which had occupied him for so long, Friedmann extended change to the Universe as a whole. Can't a homogeneous and isotropic Universe be dynamic, with a time-dependent geometry? Friedmann realized that as soon as the restriction on a static distribution of matter is lifted, the Universe becomes dynamic; if the average distribution of matter changes in a uniform way, so does the Universe.

In 1922, Friedmann presented his remarkable results in a paper entitled *On the Curvature of Space*. Friedmann showed that with or without a cosmological constant, Einstein's equations have solutions that represent a dynamic Universe. More than that, Friedmann's universes exhibit several possible types of behavior, depending on how much matter there is, or if the cosmological constant is present or not. Neglecting details that are not so important for us, Friedmann distinguished two main types of cosmological solutions, expanding and oscillating. Expanding solutions represent universes where distances between two points are always increasing. De Sitter's solution represents an extreme case where the amount of matter is so small that it can be neglected, and the cosmological constant completely dominates the expansion. For expanding solutions, with or without a cosmological constant, the presence of matter slows down the expansion, but not enough to reverse it. However, if the average density of matter is larger than a certain value, called the *critical density*, the gravitational attraction exerted by matter on itself will eventually reverse the expansion, and the Universe will collapse. In principle, this cycle of expansion-contraction can repeat itself throughout time, giving rise to the oscillating solution.

With Friedmann's work, the number of "desktop universes" proliferated, as physicists eagerly awaited clues from astronomy, hoping that one of the models would correspond to our Universe. It is ironic (but also inspiring) that today, some seventy-plus years after Friedmann, we still can't decide among a variety of possible models which one best describes our Universe. The choice has been considerably narrowed, but the question has not been resolved.

Einstein was not pleased with the possibility of dynamic universes. He wrote to Friedmann that his expanding solutions were wrong due to a mathematical mistake. Soon afterward he (and others) realized the mistake was his, and Einstein published an apology in the same journal where Friedmann's paper had appeared, calling the new expanding solutions "clarifying." He later would admit that the inclusion of the cosmological constant in the original equations of general relativity was "his biggest blunder."

Static since prescientific creation myths, the Universe was suddenly awakened from its deep slumber by the courage and brilliance of a relatively unknown mathematician. In its renewed rhythm of expansion and contraction, full of rich and ancient imagery, the dancing Universe would inspire the creativity of those who wished to understand its secrets.

RECEDING HORIZONS

Static universes, expanding universes, oscillating universes; open universes, flat universes, closed (but boundless) universes. Models proliferate, possibilities abound, and the many answers only entice even more the imagination of the desktop cosmologists, inventors of universes. Everyone is confused (but excited), from scientists to the public. So what is it going to be? After all, this is no small question. It's everybody's Universe, and we would like to know it better. Before it was simple, as religion provided answers about the Universe. Different religions, different answers. But they were answers you could *believe* in. Right or wrong was not an issue. And now, what a mess, scientists want to tell us about the Universe. Should we believe them, too?

The answer is no. You don't have to *believe* in scientists. You have to understand their ideas. Science is not a system of belief, but a system of knowledge that is developed to order the world around us. Some people may be perfectly happy with a purely scientific approach to the world around (and within) them. Others may not, as they embrace a chosen system of belief. The point is that this is no replacement game. What is essential is to avoid crossing boundaries either way, in an attempt to rule one or the other out. If we choose to cross the boundaries between science and religion, it should be in search of complementarity, as the lives of Kepler, Newton, or Einstein so powerfully remind us. In my view, personal choices of self-fulfillment should tap from both knowledge and belief. Now, back to cosmology.

It was time to forget about mathematical models of the Universe for a while and just take a good look at the sky. The answers were

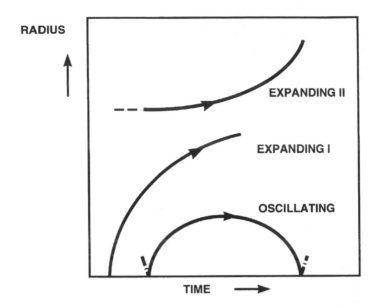

Λ	CURVATURE		
	OPEN	**FLAT**	**CLOSED**
NEGATIVE	OSCILLATING	OSCILLATING	OSCILLATING
ZERO	EXPANDING I	EXPANDING I	OSCILLATING
POSITIVE	EXPANDING I	EXPANDING I	**ALL THREE**

Nonstationary universes: The graph shows the possible solutions found by Friedmann. Expanding I starts from a zero
radius and expands forever, while Expanding II starts at a finite radius and expands forever. Oscillating solutions reach
a maximum radius and contract back to zero radius, in a re-
peatable cycle. Each of the solutions represents a family of
possibilities. The table shows possible solutions for different
values of the cosmological constant Λ.

out there, waiting. But to take a good look, you needed good instruments. To probe deeper into the night sky was to search for fainter

light sources. From vast astronomical distances, even a huge galaxy with billions of stars may appear to a powerful telescope as a very faint glow, if at all. To gather this elusive light most efficiently you need big mirrors. But big mirrors are expensive and very hard to grind. Big telescopes mean big money and advanced technology, the signatures of big science, expensive projects that involve a large number of people. Before the days of the cold war, NASA, and powerful particle accelerators, astronomy was the leading example of big science. During the first two decades of the twentieth century, through the combined action of a few entrepreneurial astronomers and the generosity of rich patrons, the focus on big astronomy moved from Europe to the United States.

George Hale perhaps best symbolizes this new approach to big astronomy. In the 1890s he convinced millionaire Charles T. Yerkes to finance a large observatory at Williams Bay, Wisconsin, operated by the University of Chicago. When completed, Yerkes Observatory had a powerful forty-inch refractor telescope, still the largest of its kind in the world, and a fairly large supporting staff.* Most people would be quite content with an observatory this size (well worth a visit if you pass by Lake Geneva, Wisconsin). But not Hale. He wanted bigger telescopes and bigger staffs, and set off to find his next patron. When Andrew Carnegie created the Carnegie Institution in 1902, Hale knew where to go for the money. He gathered enough political power to convince Carnegie to finance two new reflector telescopes at Mount Wilson in California, a sixty-inch and hundred-inch giant.

In 1917, after a long struggle that involved several technical problems ranging from grinding the huge mirror to mules that refused to climb the treacherous paths up the mountain, the hundred-inch telescope was ready to point its giant eye to the sky. That same year, a young astronomer named Edwin Hubble completed his Ph.D. at Yerkes Observatory. His thesis, "Photographic Investigations of Faint Nebulae," was not particularly distinguished, but was good enough to set Hubble on his professional course. He was interested in distant nebulae then and would remain so for most of his life. The

---

*A refractor telescope has as its principal focusing element a lens, while a reflector telescope has as its principal focusing element a mirror. "Forty-inch" refers to the diameter of the principal lens for a refractor or the mirror for a reflector.

new giant telescope and the young astronomer were destined to forge a partnership that would change the course of modern cosmology. The Universe was never the same after Hubble looked at it.

Hubble was one of those people with the golden touch: an impressive athlete, excelling at boxing, track, swimming, and basketball, a Rhodes scholar, handsome, self-assured, and a born leader. He knew what he wanted and how to get it. And, perhaps most importantly, once he got what he wanted, he knew how to tell everyone about it. He was clearly destined for greatness. However, before Hubble and the hundred-inch giant could become partners, America entered World War I. Always ready for action, Hubble eagerly joined the American Expeditionary Force destined for France. He quickly was promoted to captain and then to major. It seems that his only disappointment with the war was that he "barely got under fire." (I can't claim to know many astronomers today that would have said the same thing.) Hubble left the army in August 1919 to accept a generous job offer by Hale to join the staff at Mount Wilson. It's not hard to imagine Hubble thinking, "What's the fun of being in the army without a war to fight?" Besides, there was another war going on, this one being fought between astronomers, as opposed to armies. And there was plenty of fire in this war for Hubble to enjoy himself.

By 1920, the debate about the nature of the nebulae had reached a climax. The two competing views, that nebulae were part of the Milky Way and, instead, other "island universes" outside our galaxy, seemed to have equally convincing support, creating a very confusing situation. On April 26, 1920, Harlow Shapley from Mount Wilson and Heber Curtis from Allegheny Observatory in Pittsburgh met before the National Academy of Sciences to discuss the evidence for and against island universes. This meeting became known as "the Great Debate." Shapley was convinced that the Milky Way was much larger than believed at the time, and hence that it encompassed other nebulae. Curtis held the opposing view, that island universes were galaxies like the Milky Way but far away from it. They discussed the evidence, trying to bias it toward their own personal views with varying degrees of success.* Although Shapley's concluding remarks carried more

*So much for the belief that science always progresses in sure steps, a belief that only serves to give science a reputation of being a very inhuman activity.

strength than Curtis's, the Great Debate ended inconclusively. In order to settle the issue, better estimates of the distances to the spiral nebulae were needed. This is where Hubble came in.

Measuring distances in astronomy is extremely difficult. Just think how hard it is to estimate the distance from someone holding a flashlight turned away from you in a dark night. The standard procedure is to measure the brightness of the light at a fixed distance (called its *intrinsic luminosity*), and then use the so-called *inverse-square law* to estimate the distance. Basically, the inverse-square law states that the brightness of the light drops with the square of the distance. Now try doing this for stars, or even worse, for nebulae. We have to proceed by incremental steps, starting with the radius of the Earth, the Earth-Moon distance, the Earth-Sun distance, distance to the nearest stars by parallax, and so on, hoping that we find what is called a "standard candle" on the way, that is, an object whose brightness is constant when measured from any distance, like your flashlight with good batteries placed in different locations. The problem with measuring distances to spiral nebulae was that no standard candles could be identified. It's hard to see flashlights from a large distance. Unless, of course, you have a giant telescope at your disposal.

Late in 1923 Hubble pointed the hundred-inch to the Andromeda nebula, in search of a potential standard candle. After exposing a photographic plate for about a half hour, Hubble identified the pointlike glare of a bright star. That was a good start. With great patience (an important personality trait of astronomers), Hubble kept taking pictures of Andromeda, always searching for telltale clues that would help him determine its distance to the Sun. To his surprise and delight, he noticed that the bright star he had spotted in his first picture was no ordinary star. Its luminosity changed periodically in time, in an ordered and predictable fashion. This was the typical signature of a kind of star called "Cepheid variable," which had been extensively studied by Henrietta Leavitt from Harvard about ten years earlier.

Leavitt performed a detailed study of Cepheid variables in the Milky Way and in the Magellanic Clouds, our small satellite galaxies. After a painstaking investigation involving thousands of stars, she concluded that there was a clear relationship between the period it took for Cepheids to go from bright to dim and their intrinsic luminosity; although the periods varied from a couple of days to a couple of months, the brighter stars had consistently longer periods. As Shapley had shown in 1918, once the brightness of each Cepheid variable was corrected for the distance by the inverse-square law, they all fell approximately on the same curve. Thus, Cepheids could be used as standard candles for distance measurements. Once you spotted a Cepheid, "all" you had to do was to measure its brightness variation period and, from Shapley's curve, extract its intrinsic luminosity. Knowing the star's luminosity, you could tell its distance by the inverse-square law.

Hubble furiously hunted for Cepheids in Andromeda and other spiral nebulae. As he wrote to Shapley early in 1924,

> You will be interested to hear that I have found a Cepheid variable in the Andromeda Nebula. . . . I have a feeling that more variables will be found by careful examination of long exposures. Altogether the next season should be a merry one and will be met with due form and ceremony.

After reading the letter Shapley remarked, "Here is the letter that destroyed my Universe." The ardent defender of the "Milky Way Universe" had provided the tool that destroyed his own views.

By the end of 1924, Hubble had found twelve Cepheid variables in Andromeda and twenty-two in another spiral nebula. The Great Debate was finally over, after centuries of speculation. We live in a huge Universe, our Milky Way being just one among countless other galaxies spread across the empty vastness of space. Our place in the Universe suddenly became very mundane indeed. Our planet is not in a special place in the solar system, our Sun is not in a special place in our galaxy, and our galaxy is not in a special place in the Universe. What we do have that is special is the ability to wonder about ourselves and the world around us.

Enlarging our view of the Universe was but a first step in Hubble's assault on observational cosmology. Another grand question was hanging in the air, fueled on the theoretical side by de Sitter's model of an expanding empty Universe, and on the observational side by Vesto Slipher's measurements of redshifts of spiral nebulae.* De Sitter's model predicted a linear velocity-distance relationship between two points in space. Although it was clear to everyone that our Universe was not empty, it was expected that this relationship, or something close to it, held in a more realistic model. Some observers tried to establish the velocity-distance relation as predicted by de Sitter, but the use of faulty distance measurements discredited their results. In 1924, the German astronomer Carl W. Wirtz made the simplifying assumption that all nebulae were the same size; if this were true he could measure the distance to faraway nebulae by comparing their diameters to nearby nebulae, as if you placed the same coin at different distances from your eyes. With this, he showed that the nebulae were receding from us with velocities that increased with the distance. The problem was that nebulae are not all the same size, and Wirtz's results were not taken very seriously. Again, good distance measurements were the key to success.

Using his hundred-inch partner, Hubble and his collaborator Milton LaSalle Humason looked for Cepheids in nearby nebulae in an effort to establish their distances from us. Humason was one of those people that disprove all theories about structured education. A high-school dropout after only four days, Humason found work as a muleteer during the construction of the Mount Wilson giant telescopes. He grew very fond of the place and of astronomy (and, it seems, of the daughter of one of the engineers) and managed to get a job as the observatory's janitor once the telescopes were ready. The staff astronomers were quick to notice that Humason had a kind of magic touch with the telescopes, and they often came to him when things got out of hand. Meticulous and very patient (of course), Humason would manipulate the giant telescopes with disarming grace. Soon Humason had mastered most of the techniques of observational astronomy. To avoid further embarrassment to the other staff astronomers, by 1919 Hale found it wise to promote Humason to assistant astronomer. From then on he was allowed to make his own observations.

---

*Partly because of Friedmann's tragic premature death in 1925, his work in this area was to remain largely unnoticed until the early thirties.

In order to check for a velocity-distance relationship, reliable measurements of both velocities and distances were necessary, making it a two-step process. To check for the velocities, Hubble and Humason used the Doppler effect, searching for redshifts in the spectra of distant nebulae; the amount of redshift in the spectral lines of the nebula is proportional to its receding velocity from us. By 1929, using also some data from Slipher, they had measured the spectral shift of forty-six nebulae. And they found a clear trend toward the red.

As for the distances, initially Hubble used the same Cepheid variables technique that had put an end to the debate on island universes. However, even with the hundred-inch giant, Cepheids could only be seen in nearby nebulae. If the receding velocity indeed increased with distance, nearby nebulae would not recede as fast as far away ones, and this limitation complicated things somewhat. Local velocities in arbitrary directions, caused by the gravitational attraction between nearby nebulae, would overshadow the receding velocity, seriously undermining the results. Although the Cepheids were a reliable first step, Hubble had to find a different "standard candle" to tell the distances to faraway nebulae.

Hubble started by searching for the brightest stars he could find in nearby nebulae. After all, to be seen from a huge distance, "standard candles" had better be as bright as possible. Since he knew the distances to these nebulae using Cepheids, he could determine the intrinsic luminosity of the bright stars using the inverse-square rule. Hubble found that all the bright stars had similar intrinsic luminosities, somewhat like flashlights of the same make and model placed at different distances from us in a dark night. (He really had the golden touch.) Thus, all he had to do was to assume that the brightest stars in distant nebulae had the same approximate intrinsic luminosity as in nearby nebulae, that is, the same flashlights everywhere. If that were true, the brightest stars could be used as the new standard candles, their distances computable from the inverse-square rule.*

In 1929, Hubble wrote a paper entitled "A Relation Between Dis-

---

*This assumption doesn't hold for very faraway galaxies, as they will be at different stages of evolution and thus have bright stars with different top luminosities. However, it was a fair approximation for some of the galaxies Hubble was looking at.

tance and Radial Velocity Among Extra-Galactic Nebulae." In it, loud and clear, was his proposal that there was indeed a linear relationship between the distance and the velocities of distant nebulae. The Universe was not only much larger than was anticipated, but was also dynamic. As Hubble stated in the conclusion of his paper, a new era for cosmology had begun, linking theory to observations:

> The outstanding feature, however, is the possibility that the velocity-distance relation may represent the de Sitter effect, and hence that numerical data may be introduced into discussions of the general curvature of space.

Hubble continued to improve his data to test the velocity-distance relation to deeper distances. In 1931, he published a joint paper with Humason further improving his 1929 results. Typically concerned that he get proper credit for his discovery, he wrote to de Sitter:

> The possibility of a velocity-distance relation among nebulae has been in the air for years—you, I believe, were the first to mention it. But our preliminary note in 1929 was the first presentation of the data ... to establish the relation. In that note, moreover, we announced a program of observations for the purpose of testing the relation at greater distances—over the full range of the 100-inch, in fact. The work has been arduous but we feel repaid since the results have steadily confirmed the earlier relation. For these reasons I consider the velocity-distance relation, its formulation, testing, and confirmation, as a Mount Wilson contribution and I am deeply concerned in its recognition as such.

Two months later de Sitter and Einstein visited Hubble in Pasadena, where they jointly proclaimed Hubble's laurels. Einstein finally accepted the reality of the expansion of the Universe and permanently dropped the cosmological constant from his equations of general relativity. Despite previous hints at the existence of the velocity-distance relation, Hubble deserves the credit for its detailed formulation and confirmation from his meticulous observations. The relation, written simply as $v = Hd$, is now called *Hubble's law*, and the constant H, with dimensions of inverse time, is called Hubble's constant. If others felt they also deserved some credit for this breakthrough, nobody came forward to challenge the amateur boxer.

The plasticity of spacetime, the cornerstone of general relativity, is most wonderfully embodied in the expansion of the Universe. Drifting away from each other, carried along by the stretching geometry, billions of galaxies fill the ever-growing emptiness of space with their richness of shape and light. At all scales, from the smallest components of matter to extra-galactic astronomy, the Universe emerges as a dynamic entity, dancing to the music of transformation. A picture of motion and change becomes the modern view of nature, replacing the more rigid classical framework.

New ideas prompt new questions, and the more we discover about the Universe, the more we want to know. This never-ending curiosity is the backbone of science. Now that it was clear the Universe was expanding, cosmologists wanted to know its history. The age-old questions came back full force, to haunt and inspire. Did the Universe have a beginning? How big is it? How old is it? If it did have a "Beginning," can we understand it? How did it get from "there" to "here"? Will it have an end? As mythmakers have done since time immemorial, scientists would tackle these questions with renewed passion and devotion. Armed with their new tools of discovery, they would stretch the possibilities of scientific inquiry to its limits. Maybe even a little beyond its limits. But then, if we don't stretch our limits we can't enlarge our boundaries, and risk is curiosity's best friend. The next generation of cosmological models would merge the boundaries of the big and small, feeding from ideas coming from nuclear and particle physics. Curiosity would be rewarded in a most spectacular way, as cosmologists successfully deciphered the history of the Universe's infancy.

And yet, when it comes to questions of beginnings, a subject we also will return to next, scientific tools hit their limits of validity. Models abound, drawing from a combination of physical reasoning and personal expectations of how things should have been. As we will see, some ideas, cloaked in scientific jargon, curiously reflect mythic images put forward a long time ago, within a very different context. Our creativity seems bound to repeat itself even if dressed in different clothes. Is that a weakness of science, or, more generally, of human creativity? I really don't think so. More than anything, it reveals the common roots of human imagination, and how they are reflected in the manifold ways by which we try to make sense of the world around us. As the character Hannah in Tom Stoppard's play *Arcadia* remarks, "Comparing what we are looking for misses the point. It is wanting to know that makes us matter."

10

BEGINNINGS

Only He who is its overseer in highest heaven knows.
He only knows, or perhaps He does not know!
— RIG VEDA X

I recently rediscovered the principle of dialectics, as encapsulated in a short dialogue between me and my son Andrew:

> *Andrew (age seven):* "Dad, can anything move faster than the speed of light?"
> *MG:* "No."
> *Andrew:* "What about the speed of darkness?"

Leave it to children to remind us of the many ways to perceive reality! There is little doubt that when we attempt to order the world around us, thinking in terms of opposites is very useful. Day-night, female-male, dead-alive, left-right, rich-poor, polarities are everywhere. It may very well be that our brain itself is the product of this polarized reality, well adapted to the world within which it functions; we order the world around us in terms of opposites because our brain, being the optimized product of interactions with this external reality, is "wired" to function this way. This is a pedestrian view of our brain as the product of evolution dictated by natural selection. But then, an unpleasant question arises. If our brain, and thus the way we think, is the product of the environment within which it functions, can we ever construct an unbiased picture of the world? Can we transcend the inherent limitation of being the product of selective interactions with the environment in order to achieve some sort of all-encompassing view of reality? Or are we stuck within predetermined frames of thought? Disturbing, isn't it, to think that we are fairly limited in the ways we can grasp reality.

Whenever thoughts like these start to disturb my peace of mind, I go for a hike in the mountains or listen to music, preferably Mahler. Beauty external to us and beauty created by (some of) us. It is not long before my fears of being forever condemned to having a limited perception of the world are dispelled by the beauty of the scenery, or by the music that makes my brain pulsate anew with energy. I realize that even though horizons may exist, they are receding horizons, never to be reached. In a land of receding boundaries an eager traveler will always discover new wonders. This is my metaphor for human creativity.

And so it is that armed with our finite brain we wonder about the infinite, and about questions that transcend the bipolar reality in which we live. Of all questions that we can ask about nature, none is more fundamental than the origin of the Universe, what I called "The Question" in Chapter 1.* With the development of physical cosmology during the first three decades of this century, that is, with the promotion of cosmology to a physical science, it became possible, for the first time in the history of humankind, to address questions of origins in a quantitative way. As we will see, the laws of physics coupled to a sound observational program can actually be used to reconstruct the main features of the history of the Universe with remarkable accuracy.

Of course, this reconstruction is far from being completed (can it ever?), leaving some very important questions unanswered. Two of the open "origins" questions that are particularly interesting to me are the origin of matter, that is, where did the matter we and everything else are made of come from, and the origin of the Universe as a whole. Although these are two "origins" questions, they are quite different. While we can try to answer the question of

*If you read Chapter 1 a long time ago, a quick refresher may be quite useful at this point.

the origin of all matter using well-established (well, almost well-established) ideas in physics, the question of the origin of the Universe as a whole is much more complicated. Even though it is possible to use general relativity and quantum mechanics to build mathematical models that exhibit a self-consistent picture of a possible beginning, models are simply not enough to understand the question of the origin of the Universe. Since all these models *assume* the laws of physics to be valid as a tool to forge a possible beginning, they cannot possibly explain where the laws of physics themselves came from. If we simply say that the laws of physics were created with the Universe, we fall into an endless regression.

In my view, also shared by colleagues such as Paul Davies, it is the question of the origin of the laws of physics that truly deals with "the Beginning." And the answer to this question is beyond the scope of physical theories, at least as they are formulated at present. Should this stop us from trying? Certainly not! But reflecting on these issues, and on our intrinsic limitations when dealing with them, should restore a degree of humbleness that sometimes is forgotten in the "heat" of scientific speculation.

THE PRIMEVAL ATOM

An immediate consequence of Hubble's expansion law is that if the Universe is growing now, it must have been smaller before. Since the expansion of the Universe is an expansion of space, the distance between any two points was shorter in the past.* As we saw before, galaxies are dragged along with the expansion, like corks floating down a river. In fact, if we can picture the evolution of the Universe as a movie that we can play backward and forward at will (something we will be doing a lot here), playing it backward would imply that at some point in the past, a finite time ago, galaxies must have been lumped together in a small region of space.

---

*Gravitationally bound objects, such as galaxies and solar systems, do not expand with the Universe. The same is true of objects that are not bound by gravity, such as people and houses. Otherwise, the world would be a very strange place, even for Alice and the Red King.

It is very tempting to imagine that since we now see galaxies receding from the Milky Way in all directions, playing the movie backward in time implies that some time ago all galaxies were crowding around us. Does this mean that we are the center of the Universe? Absolutely not! Remember that the Universe has no center, that all spatial points are equivalent. What we see from our perfectly mundane spot in the Universe is what other observers would see from any other point in the Universe. Playing the movie backward, "they" would see all galaxies approaching "them" to a final crunch a finite time ago.

From the velocity-distance relation, and assuming that the receding velocities were the same throughout the expansion, Hubble extracted how long it took for the galaxies he observed to have moved from a point of initial concentration to their present distances from us.* That is, he measured the age of the Universe. His answer was two billion years. A fascinating result, although it immediately presented cosmologists with a big problem. It was known by then from geological dating techniques that the Earth was at least three billion years old! How could the Earth be older than the Universe? This obvious discrepancy didn't do any good for the credibility of cosmology.

Cosmologists initially tried to get around this problem by redefining what was meant by the age of the Universe. Maybe cosmological time and geological time are not the same thing, or cosmological time starts a little later. De Sitter was very disturbed by this situation, writing in 1932 that this was a "very serious difficulty presented by the theory of the expanding Universe," a "paradox," and a "dilemma." He even suggested, in a somewhat desperate tone, that since

> the "Universe" is an hypothesis, like the atom, [it] must be allowed the freedom to have properties and to do things which would be contradictory and impossible for a finite material structure.

I can almost see Einstein's expression of disgust with this sort of attitude. As a final plea, de Sitter suggested that maybe the assumption of homogeneity and isotropy of the Universe would have to be

---

*The Universe hasn't grown much since the 1930s, at least in terms of cosmological distance scales.

dropped in the future. It never occurred to him that perhaps Hubble's measurements were not as accurate as they appeared to be. Only in 1952 would Walter Baade show that better distance measurements lead to a Universe comfortably older than the Earth. And even today the issue is far from over. Due to severe difficulties in measuring intergalactic distances, the precise determination of the age of the Universe is still one of the hottest points of contest in astronomy. Numbers fluctuate between ten and twenty billion years. (My personal choice is somewhere close to fifteen, but personal choices don't, or at least shouldn't, help decide scientific questions.)

At about the same time that these discussions were going on, a new voice appeared in cosmology. Georges-Henri [Joseph Edouard] Lemaître was born at Charleroi in Belgium on July 17, 1894. After an uneventful and comfortable childhood, Lemaître graduated from high school with a somewhat unusual career plan. He wanted to be both a priest and a research physicist. Due to financial problems in his family, Lemaître's plans had to be postponed. At first he followed his father's sensible advice that if he were to earn a decent living he'd better forget his foolish dreams and join the engineering school. But when the calling is strong, there is no use trying to hide from it just for the promise of material comfort. Lemaître's career as an engineer was to be quite short-lived.

When I finished high school, I too wanted to be a physicist. Dreams of becoming a musician had been shattered, perhaps wisely, a few years before. "Better to be an efficient amateur than a struggling professional," my father argued. I remember my father's carefully built arguments opposing such a career choice, which culminated with him asking me if I really believed that someone would actually pay me to "count stars." As I tried to babble something about what physicists really do, my father charged on, with his unfailing certainty: "Brazil needs chemical engineers." I gave in. Maybe I could become the only chemical engineer in the world with an autographed picture of Einstein on my office wall. Maybe I could study the theory of relativity as an amateur, just as I did with music. My self-denial didn't (and couldn't) last long. After two years of disastrous laboratory work in chemistry I switched to physics, the happiest decision of my professional life. At the time, I remember being very scared about the whole thing. I went to visit my older (and sometimes wiser) brother, Luiz, who was hospitalized with a nasty case of hepatitis. After I exposed all the pros and cons of my dilemma, Luiz popped the real question: "Are you good enough?" "I think so," I answered, somewhat uneasily. "Then go for it." And I did.

Lemaître graduated in civil engineering in 1913 and started training as a mining engineer, despite his love for physics. Sometimes it takes a very dramatic event to change someone's course in life. In Lemaître's case, the "event" was more than four years' exposure to the horrors of World War I. When the war was over, Lemaître knew it was time to pursue his dream. By 1920, he had joined both a graduate program in mathematical physics and the Maison Saint Rombaut, an extension of the seminary of the Archdiocese of Malines, where adults were trained for priesthood, and in September of 1923, Lemaître was ordained a priest. In October, he joined Eddington's research group in Cambridge as a graduate student. After a year in Cambridge, England, Lemaître left for Cambridge, Massachusetts, where he joined Shapley's group at Harvard. There, he developed a solid foundation in both theoretical physics and astronomy, a combination that was to determine his constant efforts to link theoretical and observational aspects of cosmology.

In 1927, Lemaître wrote a paper in which he basically rediscovered Friedmann's cosmological solutions predicting an expanding Universe. In the same paper he showed that these solutions, like de Sitter's, also led to a linear velocity-distance relation for receding galaxies. However, the paper was published in an obscure journal and remained largely unnoticed. Lemaître did try to talk to Einstein about these results, but Einstein showed no interest: *"Vos calculs sont corrects, mais votre physique est abominable."* (Your calculations are correct, but your physics is abominable.) In spite of Einstein's remarks, Lemaître's fate was about to change dramatically. Within a few years, Einstein would be applauding his ideas.

When Hubble made his observations public, many cosmologists, including Eddington and de Sitter, scrambled to find a semirealistic model of the Universe that could accommodate both matter and the expansion. When Lemaître heard of their efforts, he reminded his ex-advisor that he had solved the problem in 1927. Eddington finally read the paper and managed to get a translation published in the journal *Monthly Notices of the Royal Astronomical Society*. Lemaître's prescient ideas were finally vindicated. Enjoying his newly found fame, Lemaître pressed on with his more ambitious plan: to develop a complete, even if somewhat qualitative, history of the Universe. As he wrote a few years later,

> The purpose of any cosmogonic theory is to seek out ideally simple conditions which could have initiated the world and from which, by play of recognized physical forces, that world, in all its complexity, may have resulted.

In 1931, Lemaître published a paper in the journal *Nature*, proposing his idea of the "Primeval Atom," where he described the initial evolution of the Universe in terms of the decay of an unstable nucleus, combining elements of nuclear physics with the second law of thermodynamics. He made no effort to explain where this nucleus came from. Interestingly, the primeval atom has been sometimes compared to the mythic "cosmic egg" of creation stories. In Lemaître's own words:

> This atom is conceived as having existed for an instant only, in fact, it was unstable and, as soon as it came into being, it was broken into pieces which were again broken, in their turn; among these pieces electrons, protons, alpha particles, etc., rushed out. An increase in volume resulted, the disintegration of the atom was thus accompanied by a rapid increase in the radius of space which the fragments of the primeval atom filled, always uniformly. . . .

He then described how, from this prototypical matter, gaseous clouds would eventually form and condense into clusters of nebulae. He even went as far as to propose that the debris of these "cosmic fireworks" were detectable today as "fossil rays," which he associated with cosmic rays. Little did he know that "fossil rays" indeed permeate the Universe, although they are not related to cosmic rays.

In a rough sense, Lemaître held the whole history of the Universe in his mind. His intuition was uncanny. Nevertheless, he was the first to acknowledge that this picture was only qualitative and not predictive. "Naturally, too much importance must not be attached to this description of the primeval atom, a description which will have to be modified, perhaps, when our knowledge of atomic nuclei is more perfect." These were prophetic words! Lemaître's cosmogonic vision, in a sense a cross between a creation myth and a scientific model, was to become the precursor of the modern big-bang model of cosmology.

What about Lemaître the priest? How did he reconcile his scientific view of Creation with his religious convictions? Lemaître made every effort to keep the two separate. He insisted that his primeval atom hypothesis was a scientific model, not inspired by his religious views on Creation. He felt very uncomfortable when in 1951 Pope Pius XII compared the initial state of the Universe as described scientifically with the Catholic interpretation of the Genesis. In 1958, due to the pressure of several colleagues, Lemaître felt it was time to justify his position:

> As far as I can see, such a theory remains entirely outside any metaphysical or religious question. It leaves the materialist free to deny any transcendental Being. . . . For the believer, it removes any attempt to familiarity with God. . . . It is consonant with Isaiah speaking of the Hidden God, hidden in the beginning of [Creation].

The mystery is in the eye of the beholder. Lemaître never ruled out the possibility that even the "coming into being" of the primeval atom could eventually be explained scientifically, proposing (again with uncanny prescience) that the answer may come from applying quantum mechanics to the Universe as a whole. As with so many other physicists we have encountered already, the aspect of his faith that was applied to his work was a deep reverence for nature and for the many wonders that are revealed to those who venture a little farther down the road than most.

THE UNIVERSE OF BEING

Despite the support of a few cosmologists, Lemaître's ideas were not taken very seriously for quite a while. To believe that there was such an event as the "beginning of everything," with all its religious connotations, was an idea that many found repugnant. How could a scientific theory of the Universe be based on an event that simply defied any possibility of being explained causally? And why should we assume that the laws of physics were valid at the extreme conditions that surely held in the beginning? Eddington himself, a devout Quaker, tried to get around the creation issue by proposing that "since I cannot avoid introducing this question of a beginning, it has seemed to me that the most satisfactory theory would be one which made the beginning *not too unaesthetically abrupt.*"

Accordingly, Eddington argued that if, in the beginning, matter was distributed perfectly homogeneously in a small volume, it would be impossible to distinguish between "undifferentiated sameness and nothingness." In this Universe, evolution would progress slowly through the growth of small imperfections, as opposed to Lemaître's cosmic fireworks.

Whatever the attempts to diffuse the "abruptness" of an acausal appearance of a Universe at some instant in the past, evolutionary models of cosmology suffered from another more immediate problem: Hubble had measured the Universe to be younger than the Earth.

The combination of philosophical distaste for a Universe with a beginning along with Hubble's age measurements led a trio of young British physicists to propose a completely different model for the Universe. In the so-called *steady-state* model of cosmology the Universe has basically always been the same, having no beginning or end in time. It was a Universe of being, with no creation event, reminiscent of the Jain myth of creation we examined in Chapter 1, or of Parmenides's Eon. From a philosophical point of view, the motivations that led the British trio to propose the steady-state model were not so distinct from the Jainists or the Eleatics: an aversion to a creation event and an aversion to change, respectively. Although the steady-state model is by now largely discredited (there remain a few die-hards out there, but they are very few indeed), its brief life provides us with some important pointers in the development of physical cosmology.

In 1948, Thomas Gold and Hermann Bondi, and independently, Fred Hoyle, all from Cambridge, published two papers in the *Monthly Notices* describing their new cosmological theory without a creation event. Although some of the details in the two papers are very different, they are often taken as representing the steady-state "school of thought."* They proposed an extension of Einstein's cosmological principle, called the *perfect cosmological principle*, where the Universe was not only the same everywhere but *always*. There was no age problem because the Universe was infinitely old. Both time and space were to be treated in the same way, so to speak. All they had to do was somehow to accommodate the observed recession of galaxies into their Universe of being.

As the Universe expands it "thins out," as less and less matter occupies a given volume. This thinning-out implies that the older the Universe is the less dense it gets, the trademark of any evolutionary cosmology. However, in the steady-state model the Universe cannot thin out, as this represents change. In order to get around this effect, Bondi, Gold, and Hoyle proposed that as the Universe expanded and thus thinned out, more matter was created to "fill in" the gaps in such a way that the overall matter density remained constant. An analogy may help. Imagine you filled up your bathtub with water. Now pull the plug and let the water go down the drain. You can measure how fast the water is going down the drain by following the water line on the bathtub. If you turn on the faucet in such a way that the exact amount of water that is being drained is also being poured back into the bathtub, you will achieve a steady-state situation; as long as your water supply lasts, the level of the water in the bathtub will remain constant.

"Wait a minute," you exclaim. "It's okay to fool around with water in my bathtub, but where is this extra matter coming from to fill in the gaps in the steady-state model?" Very good question. Spontaneous creation of matter violates our dearest law of all, conservation of energy. Of course the British trio was well aware of this fact. They astutely replied that we can only infer that energy is conserved by making measurements. Since every measurement has a limited accuracy to it, how do we know if energy is truly *exactly* conserved?

---

*Legend has it that the core idea of the steady-state model came to Gold when he, Bondi, and Hoyle went to a movie about a bizarre ghost story, which ended as it started. The absence of a clear time evolution in the movie's plot prompted Gold to ask his friends if the same couldn't be true for the Universe.

When you quantify how much matter must be spontaneously created to keep the Universe in a steady-state, you come up with the absurdly small rate of about three atoms of hydrogen per cubic meter per million years. No one can possibly measure a violation of energy conservation at this level. Also, the trio would argue, is continuous creation of matter any worse a violation of this principle than the abrupt creation of the Universe?

As we'll soon see, at about the same time that the steady-state model appeared in the press, the Russian-born George Gamow and his collaborators were proposing the big-bang model. None of the two teams or their defenders took the other seriously. While the British claimed that the big-bang model was philosophically indefensible, Gamow claimed that the steady-state was too philosophical. In fact, the term "big-bang" was coined by Hoyle as a derisory comment on the idea of a beginning. With the (rarely) courteous civility that characterizes scientific disputes, the two teams agreed that the final word on the models lay beyond philosophical prejudices and was to be decided by observations.

The first blow to the steady-state model came around 1952, when Walter Baade, using the 200-inch Hale telescope at Mount Palomar, showed that Hubble had made a mistake in his distance measurements. The age of the Universe immediately doubled and soon grew by a factor of at least five, old enough to accommodate geological evidence on the age of the Earth. There was (at least temporarily) no age problem. The second blow came in 1955, when Cambridge radio astronomers led by Martin Ryle (opposition at home is always the toughest!) showed that their survey of radio sources (astrophysical objects that generate electromagnetic radiation in the radio end of the spectrum) contradicted Hoyle's calculations; the steady-state model predicted fewer sources than what was found by the Cambridge survey.

The final blow came in 1965, when it was discovered that the whole Universe is permeated by a background of very cold photons. As we will see next, this background had actually been predicted by the proponents of the big-bang model as being the "fossil rays" from an earlier era when the Universe was much hotter. The steady-state model could not offer any plausible explanation for the existence of this background and had to be abandoned as a serious cosmological theory. Change and transformation characterize the physical Universe. As Heraclitus so beautifully expressed over twenty-five centuries ago, "You can never step in the same river twice."

THE UNIVERSE OF BECOMING

One of the most remarkable feats of Lemaître's "primeval atom" hypothesis was the suggestion that the early evolution of the Universe was strongly dependent on the physics of the very small, namely atomic and nuclear physics. Lemaître's vision suggested a deep connection between the physics of the very small and the physics of the very large. Granted, when Lemaître proposed his radioactive "cosmic egg" in 1931, not much was known about the physics of the nucleus. The companion to the proton in the nucleus, the neutron, was only discovered in 1932. But the thirties saw a very rapid development of nuclear physics, which, as we all know, culminated in the dramatic bombing of Hiroshima and Nagasaki in 1945.

The unearthly release of energy that occurs as heavy radioactive nuclei are split into smaller ones by being bombarded with neutrons marked a radical shift in the collective history of humanity; for the first time ever, we had the power to completely annihilate ourselves many times over. Oppenheimer remarked after the successful detonation of the first atomic bomb in the desert of New Mexico:

> We waited until the blast had passed, walked out of the shelter, and then it was extremely solemn. We knew the world would not be the same. A few people laughed, a few people cried. Most people were silent. I remembered the line from the Hindu scripture, the *Bhagavad-Gita*: Vishnu is trying to persuade the Prince that he should do his duty and to impress him he takes on his multi-armed form and says, "Now I am become Death, the destroyer of worlds." I supposed we all thought that, one way or another.

Leaving aside the somber military uses of nuclear energy, the 1930s also saw the application of nuclear physics to the study of astrophysical objects. In particular, it became clear that the enormous energy that continuously pours out of stars is the result of a chain of nuclear reactions. By the late thirties, the German-born physicist Hans Bethe and others had developed a theory that explained why the Sun shines. It also became clear that the reactions that make the Sun shine are very different from the reactions that power the bomb. While the fission reactions that power an atomic bomb or a nuclear power station are in a sense "destructive reactions," with energy being released as large nuclei are split into smaller ones, the fusion reactions powering the Sun are "constructive reactions," a progressive buildup of larger nuclei from smaller ones.

Since this concept of progressive buildup of larger nuclei from smaller ones is of extreme importance not only for understanding why stars shine, but also for understanding the early history of the Universe, perhaps we should take a few minutes now to remind ourselves of how chemists organize the table of elements.

There are ninety-two naturally occurring chemical elements. How do we distinguish between them? By the number of protons they have in their nucleus. Thus, hydrogen, the lightest element, has one proton in the nucleus, while helium, the next heaviest element, has two, lithium has three, and so on until uranium, which has ninety-two protons in its nucleus. Because protons have a positive electric charge, an atom that is electrically neutral must have the same number of protons and electrons; the total net negative charge of the electrons neutralizes the total net positive charge of the protons. Thus, neutral hydrogen has one electron, neutral helium has two, and so on. The final member of the atomic partnership is the neutron, slightly heavier than the proton and with no electric charge. (Hence the name, neutron.)

Protons and neutrons make up the nucleus, and electrons "orbit" around it.* But if we want to understand how atoms work we need to investigate which forces operate between its building blocks. We know that since the proton and the electron have opposite electric charges, they are attracted to each other by the electromagnetic force. We also know, from the uncertainty principle in quantum mechanics, why the electron doesn't fall onto the nucleus.

---

*The quotes are to remind us that the concept of orbit is not really adequate to describe the trajectories of electrons around nuclei, as we know from our discussion of quantum mechanics.

But within the nucleus there are all these protons repelling each other, and yet something keeps the nucleus together. This something is the so-called *strong force*, which is roughly one hundred times stronger than the electromagnetic repulsion the protons feel. The reason why we don't know about the strong force in everyday life is because it is a very short-range force. Contrary to gravity and electromagnetism, which have an infinite range decaying in strength with the square of the distance, the strong force only operates within nuclear distances.

The fascinating thing about the strong force is that it not only binds protons to each other, but it also binds neutrons to protons and to each other. Hence, a nucleus with many protons and neutrons is bound together by the strong force. The fact that the strong force does not care about electric charge adds an interesting twist to nuclear physics. Since neutrons are electrically neutral, it is possible for a given element to have different numbers of neutrons in its nucleus. For example, a hydrogen atom is made of a proton and an electron. But it is possible to add one or two neutrons to its nucleus. These heavy "cousins" of hydrogen are called *isotopes*. The hydrogen isotope deuterium has a proton and a neutron, while tritium has a proton and two neutrons. Every element has several isotopes, which, as with hydrogen, are built by adding (or extracting) extra neutrons to their nuclei.

Now that we know about nuclei and isotopes, we can go back to the idea of progressive buildup. How do we synthesize deuterium? By fusing a proton and a neutron. What about tritium? By fusing an extra neutron to deuterium. And helium? By fusing two protons and two neutrons, which can be done in a variety of ways.* And the buildup continues, as heavier and heavier elements are synthesized inside stars, up to the element iron. Whenever a fusion process occurs, energy is released, called the *binding energy*. It is identical to the energy we must provide to the bound system of particles to break their bond. This concept of binding energy is also useful outside nuclear physics. Any system of particles, not only protons and neutrons, that are bound by some force has a binding energy associated with it. For example, a hydrogen atom, made of a bound

---

*For example, by fusing two deuterium isotopes, or a tritium and a proton.

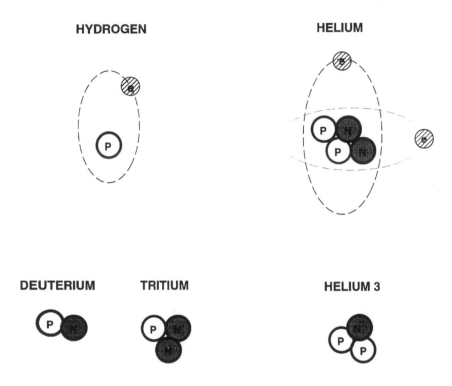

Atoms and isotopes. Each element has a characteristic
number of protons, while its isotopes are formed by adding
or extracting neutrons.

proton and an electron, has a binding energy. If I disturb the atom
by giving it a "kick" with energy larger than its binding energy, I will
break the bond between the proton and the electron, which will
then move freely away from each other. What Bethe and others
realized was that the stars were like alchemical laboratories, which
could transmute lighter elements into heavier ones, releasing a
tremendous amount of energy in the process. This buildup of heavier nuclei from smaller ones is known as *nucleosynthesis.*

However, any energy-producing system must run out of fuel at
some point. A star consumes itself in order to exist, its life a desperate struggle between gravity pulling the star in and nuclear energy
pushing it out. The star's existence depends on the delicate balance
between these two opposite trends. Originally made mostly of hy-

drogen, as the burning continues for billions of years, the star will eventually run out of fuel. Its fate will depend on its mass. For stars eight or less times as massive as the Sun, the hydrogen in the star's core burns into helium, helium burns into carbon, and carbon into oxygen. The tremendous release of energy from these fusion processes thrusts the material in the outer layers of the star into outer space, creating a planetary nebula. For stars heavier than eight times our Sun, the enormous pressure due to gravity will fuse even heavier elements at the core all the way to iron, the most tightly bound nucleus. The star then explodes with tremendous fury, in what is known as a *supernova* explosion. The carbon, oxygen, and other heavier elements that are not only part of us but essential to our survival were once synthesized inside dying stars and then spread out into space by a supernova explosion. In that sense, we are truly made of star dust.

One of the physicists working on the mechanisms by which stars burn was George Gamow, the future pioneer of the big bang model. Born in Odessa on March 4, 1904, Gamow's graduate advisor at the University of Leningrad (now Saint Petersburg) was none other than Alexander Friedmann, the man who first proposed a dynamic Universe. With Friedmann's tragic early death in 1925, Gamow became what could be called an academic orphan. After completing his thesis, he spent some time in Copenhagen and Cambridge, becoming a professor in Leningrad in 1931. For the next two years, Gamow tried several times to leave the Soviet Union, where he found it impossible to work. Party philosophy had invaded academia, dictating what should or should not be studied by scientists. For example, cosmology of any sort, or anything related to Einstein's theory of relativity, was condemned as being against dialectic materialism. Dogmatism necessarily leads to ignorance. Ignorance invites dogmatism.

In 1933 Gamow managed to escape while attending a conference in Belgium. Soon after, he accepted a professor position at George Washington University in Washington, D.C., where he stayed until 1956. In 1935, Gamow published his first paper on stellar nucleosynthesis, *i.e*, the synthesis of nuclei in stars. This work prompted an obvious question. Could the fusion processes occurring inside stars explain the synthesis of *all* elements in nature?

Furthermore, could it explain why certain elements are more abundant than others? By 1946, Gamow was convinced that stellar nucleosynthesis was not enough to explain the abundances of all elements, especially the lightest ones. He then boldly explored another possibility: The lightest elements were cooked during the first instants after the beginning of time. Gamow's proposal turned the early Universe into a cosmic furnace.

A hot early Universe was not a new idea. Although Lemaître's primeval atom was sufficiently cold to preserve nuclear bonds, during the 1930s Richard Tolman from the California Institute of Technology and others realized that a dynamic Universe that followed Friedmann's expanding model was necessarily very hot and very dense in its earliest stages of evolution. Without dealing with the more complicated issue of the "initial singularity," that is, the point from which space and time itself originated and the laws of physics as we know them break down, Tolman applied thermodynamics to the expanding Universe to show that it cooled as it evolved from its initial primeval fireball stage. Furthermore, it was possible to predict the temperature of the Universe at a given stage of its evolution. All that was needed was the amount and composition of matter and of radiation (that is, photons) constituting the "primordial soup." If plausible assumptions could be made about these ingredients, a history of the Universe could be reconstructed practically from scratch.

In 1947, Gamow enlisted the help of two collaborators, Ralph Alpher and Robert Herman. Alpher was then a graduate student at George Washington University, while Herman worked at the Johns Hopkins Applied Physics Laboratory. Within the following six years, the three would develop the physics of the big-bang model pretty much as we know it today. Gamow's picture starts with the Universe filled with protons, neutrons, and electrons. This is the "matter component," which Alpher called *"ylem."* Added to that were very energetic photons, the "heat" of the early Universe. The Universe was so hot at this early time that no binding was possible; every time a proton tried to bind with a neutron to make a deuterium nucleus, a photon would come racing to hit the two away from each other. Electrons, which are bound to protons by the much weaker electromagnetic force, didn't have a chance. If there's no binding when it's too hot, what temperatures are we talking about? Around one thousand billion degrees Fahrenheit. The image of a cosmic (or primordial) soup is an appealing way to describe these very early stages in the history of the Universe. The different building blocks of matter just roamed freely, colliding with each other and with photons, but not binding to form nuclei or atoms, somewhat like floating vegetables in a hot minestrone soup. As the big-bang model evolved to its accepted form today, the basic ingredients of the cosmic soup changed somewhat, but not the basic recipe.

From this initial state, structure started to emerge. The hierarchical clustering of matter progressed steadily, as the Universe expanded and cooled. As the temperature lowered, that is, as the photons became less energetic, nuclear bonds between protons and neutrons became possible. An era known as *primordial nucleosynthesis* started, where deuterium, tritium, helium and its isotope, helium 3, and an isotope of lithium, lithium 7, were formed. The lightest nuclei were cooked in the earliest moments of existence of the Universe. This is how Gamow, in his wonderfully irreverent tone, wrote his version of Genesis:

> In the beginning God created radiation and ylem. And ylem was without shape or number, and the nucleons [protons and neutrons] were rushing madly over the face of the deep.
> And God said: "Let there be mass two." And there was mass two. And God saw deuterium, and it was good.
> And God said: "Let there be mass three." And there was mass three. And God saw tritium and tralphium,* and they were good. . . .
> And God said: "Let there be Hoyle." And there was Hoyle. And God looked at Hoyle and told him to make heavy elements in any way he pleased.
> And Hoyle decided to make heavy elements in stars, and to spread them around in supernova explosions. . . .

For Gamow and collaborators, the synthesis of light nuclei took about forty-five minutes. With more modern values for the various nuclear reaction rates involved, it took about three minutes. The remarkable feat of Gamow, Alpher, and Herman's theory was that

*Gamow's name for helium 3.

they could *predict* the abundance of these light nuclei. In other words, using relativistic cosmology and nuclear physics, they could tell how much helium should have been synthesized in the early Universe. It turns out that about twenty-four percent of the Universe is made of helium. Their predictions could then be checked against what was produced in stars and compared to observations.

Although the big-bang model attracted increased attention in the 1950s, partially due to its scientific merits and partially to Gamow's efforts to popularize his ideas, no extensive observational program was initiated to check its predictions. The idea of using the early Universe as a furnace to cook the lightest nuclei was considered a bit too exotic for serious consideration. Why not try to determine if all elements are synthesized in stars, which at least we know are there? Several models appeared in the literature trying to show that it was possible to produce as much helium from stars. However, by 1964, it became clear that far more dramatic conditions than the interior of stars were needed to generate the necessary amount of helium. To a chosen few, Gamow's ideas became very compelling.

Helium and other light nuclei synthesized in the early moments of existence of the Universe are very much like fossils from those distant times. In a sense, the work of the cosmologist is akin to that of the paleontologist, as we try to reconstruct a whole picture of an era and its subsequent evolution through the analysis of very scarce fragments or clues. The main difference is that physics has a predictive power that can make the job of cosmologists somewhat simpler; we can actually make a prediction that a certain kind of fossil should exist. We then look for the fossil and, if it's there, great, the theory is working. Otherwise, we must refine or change certain assumptions behind our modeling. The main problem, common to both cosmology and paleontology, is that some fossils are very hard to find.

Aside from the abundance of light nuclei, Gamow made a much more dramatic prediction. After the era of nucleosynthesis, the "ingredients" of the cosmic soup were mostly the light nuclei, electrons, photons, and neutrinos, particles that like photons have no mass or electric charge. The next step in the hierarchical clustering of matter is to make atoms. As the Universe expanded, it cooled, and the photons became progressively less energetic. At some point, when the Universe was about three hundred thousand years of age, the conditions were ripe for electrons to bind with protons to make hydrogen atoms. Before this time, whenever a proton and an electron tried to bind, a photon would kick them apart, a sort of unhappy love triangle that just couldn't get resolved. But as the photons cooled down to about 6,000 degrees Fahrenheit, the attraction between protons and electrons won over the interference from photons, and binding finally occurred. In the process of *decoupling*, the photons were suddenly liberated to pursue their dance across the Universe. They were not to interfere with atoms anymore, but just to exist on their own, impervious to all this binding that seems to be so important for matter.

Gamow realized that these photons would have a distribution of frequencies like that of a blackbody spectrum. Although the temperature of the spectrum was high at the time of decoupling, as the Universe has been expanding and cooling for about fifteen billion years since then, the present temperature of the photons would be very low. Earlier predictions were not very accurate, as they are sensitive to details of nuclear reactions that were not known accurately by the late forties. However, in 1948, Alpher and Herman predicted that this cosmic bath of photons would have a temperature of 5 degrees Kelvin, or 5 degrees above absolute zero, or about −451 degrees Fahrenheit.* (The current value is 2.73 degrees Kelvin.) Thus, according to the big-bang model, the Universe is a giant blackbody, immersed in a bath of very cold photons peaked at microwave wavelengths, the "fossil rays" from its hot early infancy.

Although the prediction was firm and the technology was available by the mid-fifties, no one set out to look for the background photons. Only in 1964 did a Princeton University group led by Robert Dicke decide to build a radio antenna to search for the photons. Meanwhile, not too far from Princeton, Robert Wilson and Arno Penzias, from Bell Telephone Laboratories, were using a twenty-foot radio antenna to study the radiation emitted from a supernova remnant located about ten thousand light-years from the Earth. The signal was very faint and they needed enormous accuracy

---

*Absolute degrees are measured in the Kelvin scale. To transform from Kelvin to Fahrenheit use the formula $F = 9K/5 - 459.4$, where K is the temperature in Kelvin and F is the temperature in Fahrenheit.

in their measurements. To their annoyance, a sort of background hiss was compromising their measurements, similar to when you try to listen to a distant conversation in a crowded party room. They checked and rechecked their equipment, but were unsuccessful at tracking down the origin of the hiss. Even a couple of pigeons that had comfortably nested inside the antenna were removed, together with the remains of their bodily functions. But to no avail. The hissing persisted, and as Penzias and Wilson soon discovered, it was insensitive to *where* they pointed the antenna. It was coming from all directions in the sky!

They did what scientists do when they are in trouble; they talked to colleagues to see if anyone had any idea as to why this was going on. Eventually, the trail led them to nearby Princeton, where Dicke and his group were still working on their antenna. Jim Peebles, a young theorist working with Dicke, had independently rediscovered the arguments for a background radiation of photons, the remnants of the big-bang. It all made sense now! Penzias and Wilson had discovered the "fossil rays" left over from decoupling, a snapshot of the Universe when it was a mere three hundred thousand years of age. For over ten billion years these photons have been traveling through space, the living proof of the hot beginnings of the Universe, the great triumph of the big-bang model.

The papers by Penzias and Wilson and by the Princeton group appeared side by side in an issue of the *Astrophysical Journal* in 1965. For their discovery, Penzias and Wilson won the Nobel prize in 1979. Gamow, who died in 1968, must have smiled (actually, Gamow probably jumped up and down or went for a wild motorcycle ride) when he finally saw his work vindicated. Yes, the Universe was indeed a very hot furnace that cooked the light elements and left over a background of photons that permeated space. Many physicists expressed regret for not taking the ideas of Lemaître, Gamow, Alpher, and Herman seriously long before the mid-sixties. But then, as we have seen again and again, some ideas just have to be hammered into existence before they can be widely accepted.

Cosmogony Revisited

The big-bang model, as developed by Gamow, Alpher, and Herman, reconstructed the history of the Universe from about a ten-thousandth of a second after the "bang" all the way to the decoupling of photons at three hundred thousand years, the event that originated the microwave background radiation found by Penzias and Wilson. It is indeed remarkable that relativistic cosmology and nuclear physics could combine in such a way as to provide us with a solid picture of the Universe's infancy. But of course, it is not enough. Once we have a plausible scientific scenario that reconstructs the early history of the Universe, we want to push the boundaries farther back into the past, in fact, as far back as possible. The question in everyone's mind then becomes, "Can cosmologists go all the way back to the 'Origin'? Is science really capable of providing the answer to 'the Question'?"

We touched on this issue earlier on, when I argued that mathematical models used to describe the origin of the Universe cannot be the final answer. One of them may eventually become the final *scientific* answer to "the Question," but it is not clear to me that the question of the origin of the Universe is to be answered by science alone. At least not in the way science functions nowadays. The reason for my skepticism lies in the fact that any model for the origin of the Universe must rely on accepted theories of physics, *vis-à-vis* general relativity and quantum mechanics, or possible extensions of these two.* Whenever a physicist comes up with a model of the origin of the Universe, he or she is using well-known physical laws. Thus, the building of a workable model for the origin of the Universe does not address the question of the origin of the laws of physics, or why this Universe operates the way it does. Science certainly provides many answers to nature's working, but we should not lose sight of its limitations. The question of why there is something rather than nothing should inspire us all with humbleness.

---

*There have been several books in the recent literature dealing with extensions of quantum mechanics and relativity that can be applied to the origin of the Universe, the most popular being Stephen Hawking's *A Brief History of Time*. In the Bibliography I suggest a few other titles for the interested reader.

Instead of providing a short survey of the many ideas that appeared within the last two decades addressing scientifically the question of the origin of the Universe, I will follow a somewhat less conventional route. No doubt some of these ideas will be mentioned as we go along, but my intention is not to be exhaustive or pedagogical as this would easily cover another book; I will selectively use what is needed to get my point across.

We start by going back to the first chapter of this book. As you recall, the central point of that chapter was the classification of cosmogonical myths based on how different cultures dealt with the question of Creation.

By focusing on the question of the beginning of time, I argued that there are two kinds of creation myths: myths that assume a beginning of time, and myths that assume that the Universe existed for all eternity. Within each of these two classes, I showed that there were more choices: myths with a beginning of time either imagine the world appearing as the result of the action of a creator or creators, or have the world appearing out of nothing, or as a result of order emerging spontaneously out of primordial chaos. Myths without a beginning of time fall into two classes: either the Universe is uncreated and eternal, as in the example from Jainism, or the Universe is continuously created and destroyed in a cyclic fashion, as in the example from Hinduism represented by the dancing Shiva. We should keep this classification in mind as we go along.

Now we turn our attention to the several cosmological models that resulted from applying relativity to the Universe as a whole. Can we also obtain a classification of modern cosmogonical models that follows in spirit the classification of cosmogonical myths from Chapter 1? The answer is yes. We can classify the models according to how they handle the question of the beginning, just as we did for creation myths. Again, models either assume a beginning of time, or they don't. (There is not much room for anything in between.)

Let me start with models without a beginning. We have already encountered two possible kinds of such models. One, the steady-state model proposed by Bondi, Gold, and Hoyle in 1948, assumed the Universe existed forever, with no beginning or end, and that matter is continuously created in order to keep the average energy density of the Universe constant. The other kind of cosmogonical

A CLASSIFICATION
OF
COSMOGONICAL MODELS

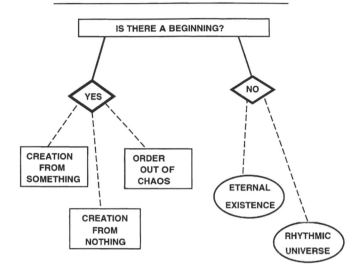

Different scientific answers to "the Question"

model with no beginning is the cyclic universe, or Phoenix universe, as it has also been called. We saw that Friedmann models with a closed geometry will lead to a universe that will have periods of expansion followed by contraction, big-bangs followed by big-crunches. Although it has been argued that due to the production of entropy within each cycle, only a small number of cycles would have been possible so far, conceptual problems related to applying general relativity and thermodynamics to the very extreme conditions close to the beginning somewhat cloud the issue. For us, the important point is that mathematically at least, such a Phoenix universe is a possibility.

What about models with a beginning? Here, also, as in the case with creation myths, the models proposed so far fall into three classes. There are models that assume a "creation out of something," "creation out of nothing," or "order out of chaos." An example of a model that assumes "creation out of something" is Lemaître's primordial atom hypothesis. He did not explain where his "cosmic egg" came from, but once it was assumed to exist, the rest followed from the laws of physics, even if only in qualitative terms. Gamow's original hot big-bang mold also assumes an initial state where certain particles of matter were present. Although it clearly differs from Lemaître's cold big-bang (nuclear bonds were not broken by heat in the primeval atom hypothesis), it does fall into the same category of assuming *something* to start with, without questioning too much (or at all) where this something came from.

Modern extensions of Gamow's model follow a similar path. With the remarkable development of particle physics during the last four decades or so, it became clear that the fundamental building blocks of matter were not protons and neutrons, but that protons, neutrons, and many other particles discovered in accelerators were actually made of yet smaller constituents, the so-called *quarks*. Since whenever there is new physics there is new cosmology, quarks were transplanted to the early history of the Universe. Before the existence of protons and neutrons, the Universe was populated by free quarks, electrons, and photons, and this allowed the clock to be pushed farther back than the ten-thousandth of a second of Alpher and Herman's work.

This tradition has continued through the 1980s and 1990s, and with new ideas coming from particle physics, the clock has been pushed back to a time of one thousand-billionth of a second, or 10^{-12} second from the bang. This time may sound preposterously small for our standards, but for the elementary particles of matter that roamed across the early Universe it is very large. For example, it takes a photon about 10^{-24} second to travel a distance equal to the radius of a proton. Moving even farther back is possible, although we leave solid physics and enter the realm of solid speculation. (I can write in this somewhat sarcastic tone because I have been involved for years in the effort of trying to push the time boundary backward.) There are many very inspired and truly beautiful ideas proposed during the past twenty years or so that must still wait for experimental confirmation before they can be considered physical. It is a bittersweet facet of research that nature doesn't give her secrets away so easily.

Mathematically, an extrapolation of any of the Friedmann models to time equals zero leads to what we call a *singularity*; the matter density becomes infinite, the spacetime curvature becomes infinite,

t < 0.00001 sec

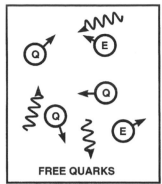

t = 0.00001 sec

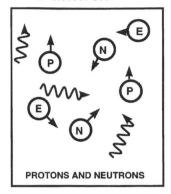

t = 1 sec

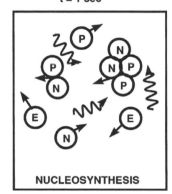

t = 300,000 years

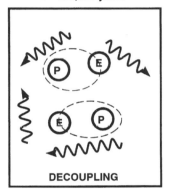

A few snapshots of the early history of the Universe. Times are approximate, as is the depiction of physical processes taking place at each frame. The wavy lines represent photons. Quarks are the constituents of protons, neutrons, and other particles that interact via the strong force.

and the distances between any two "observers" (not many around then) goes to zero. Disturbing as this may sound, the existence of the singularity is not to be taken too seriously. It signals the breakdown of general relativity and of physics as we know it at the extreme conditions that prevailed during the very first moments of existence of the Universe. That is, the singularity signals our ignorance of physics at these very high-energy scales. Something else is needed here, and ideas have not been short in coming. The most promising of these ideas call for a blend of general relativity and quantum mechanics in one form or another.

As we have seen in Chapter 8, the most dramatic effect from the Probabilistic Nature of quantum mechanics is an intrinsic fuzziness of matter that manifests itself at atomic and subatomic distances. Well, close to the singularity the whole geometry of the Universe is to be treated under quantum mechanics, and as such the very concepts of space and time become blurry. It may be that quantum mechanics will do away with the sharpness of the singularity by making it fuzzy. Attempts to marry Einstein's general relativity with quantum mechanics have been plenty, but so far have only met with more promise than success. There are enormous conceptual (and mathematical) barriers to overcome, making the task a very hard one. While some of the best minds in theoretical physics are at this moment very busy trying to perform this marriage, claims of understanding physical conditions near the singularity must be taken with a very large grain of salt. Nevertheless, we should certainly push forward and try to obtain at least some information about the very peculiar physics that dominated the beginnings of the Universe.

One interesting idea of how to apply quantum mechanics to the beginning of the Universe is the one proposed by Edward Tryon in 1973, while he was at Columbia University. Tryon made use of the fact that fuzziness in quantum mechanics is not only limited to measurements of position and velocities, but also applies to measurements of energy and time. That is, in the world of the very small, it is possible to violate the law of conservation of energy for very short times.

This is not as crazy as it seems. Think of a billiard ball lying quietly on the ground. If it is not moving, it has no kinetic energy, and if we measure gravitational potential energy from the ground up, it also has no potential energy. The ball is at a zero energy state. Now make the ball into an electron. According to Heisenberg's uncertainty principle, we cannot localize an electron and tell its velocity simultaneously. The fuzziness inherent in the electron prohibits that. Thus, in quantum mechanics, there is no zero energy state but only the lowest possible energy state of a system, its ground state. Now, if there is an inherent uncertainty in the measurement of the energy of a system, then it is possible for the energy of the ground state to fluctuate. If we call this ground state the *quantum vacuum*, we can say that the quantum vacuum always has some structure to it, that there is no such thing as a "true vacuum," in the sense of complete emptiness. In quantum mechanics nothingness is an empty concept.

If there are energy fluctuations in the quantum vacuum, very interesting things can happen. For example, from the $E = mc^2$ relation, we know that energy and matter are interconvertible. Thus, a vacuum energy fluctuation can be converted into particles of matter! Sounds absurd? Maybe, but it happens all the time. These particles are called *virtual* particles, living a fleeting existence before plunging back into the ever-busy quantum vacuum. Tryon extrapolated the idea of quantum fluctuations to the Universe as a whole. He reasoned that if, in the beginning, all that existed was a quantum vacuum, *i.e.*, "nothingness," out of this vacuum an energy fluctuation could have given rise to the Universe itself. In other words, Tryon proposed that the whole Universe was the result of a vacuum fluctuation, that it originated from "quantum nothingness." Tryon's proposal falls into the category of universes with a beginning but created out of "nothing." However, what is meant by nothingness here as well as in all other examples of quantum creation of the Universe that followed Tryon's inspiring idea must be understood in terms of "quantum mechanical nothingness" and not absolute nothingness as in complete emptiness.

Finally, there are models where the concept of singularity is substituted by a sort of geometrical chaos. The Universe as we know it, with its well-behaved isotropic geometry, emerges out of this primordial chaos as time moves forward. This idea, known as the mixmaster universe, was proposed by the American Charles Misner in 1969, and was further developed by the Russians V. A. Belinsky, I. M. Khalatnikov, and E. M. Lifshitz in 1970. Misner argued that as the Universe approached the singularity, there was no reason to assume that the geometry was isotropic, that is, that all three directions of space evolved in time the same way. By setting the geometry free, so to speak, Misner showed that a very complex behavior ensued, in which the three directions of space alternated periods of expansion and contraction, which became increasingly more vigorous as the singularity is approached. (Time is running backward here.) Thus,

in the mixmaster universe, the whole concept of singularity becomes somewhat blurred by the chaotic dance of growth and contraction of the anisotropic geometry. In their classic textbook on
general relativity, Misner, Kip Thorne, and John Wheeler wrote:

> The extrapolation of the universe's evolution back toward the sin-
> gularity at $t = 0$ therefore shows an extraordinarily complex behav-
> ior, in which similar but not precisely identical sequences of
> behavior are repeated infinitely many times.

My intention in bringing the classifications of cosmogonical
myths and cosmogonical models together should be fairly clear by
now. The examples I used here to illustrate my classification of cosmogonical models are far from being exhaustive. There is a large literature on the subject with many variations on the basic themes I
tried to address. However, I hope that the interested reader, armed
with what she or he has learned in this book, will be able to discern
within which class a given model falls, or if there is need to enlarge
this classification.

I do not claim that modern scientific theories are simply rein-
venting ancient ideas about Creation. Clearly, the language and
symbols belonging to each are completely different, and there was
most certainly no intention on the part of twentieth-century cosmologists to build mathematical models of the Universe inspired by
ancient creation myths. As I stressed elsewhere in this book, scientific models are quantitative descriptions of the natural world and
must be subjected to scrutiny before being accepted as valid, while
myths enjoy a greater creative freedom. And yet, the desire to make
sense of the Universe in which we live is common to both, as is the
awe that inspires mythmakers and cosmologists alike when con-
fronted with the most fundamental of all questions.

When we compare the two classifications, making due room for
differences of interpretation of what, say, "nothingness" or "chaos"
means in each context, we are led to an inescapable sense of repeti-
tion, that the essential metaphors behind myths and models have
many overlapping points. Although this overlap may suggest that
this comparison could be carried further, I believe that such approach would be misleading. There is not much sense in comparing

the details of the Jainist myth of creation with the steady-state model, or the Hindu myth of the dancing Shiva with the cyclic Friedmann solution.

My aim in developing these classifications and bringing them together here is to stress their common metaphors, the common mythopoetic *imagery* that mythmakers and cosmologists use when thinking about the beginning of the Universe. Together they expose both the richness and the limitations of human creativity when faced with the ultimate origin of all things. Richness for the beautiful variations on the basic themes, the versatility and color of mythic and scientific language that unfold through the many creation stories and physical models within each kind of answer to "the Question." Limitations due to the existence of only a finite number of possible answers, an ultimate bound on our fancy. We cannot but explain the existence of the Universe through our human imagination, through our many stories and models of receding horizons. Being precedes Becoming.

EPILOGUE

DANCING WITH THE UNIVERSE

While with an eye made quiet by the power
Of harmony, and the deep power of joy,
We see into the life of things.
—WILLIAM WORDSWORTH

F rom the ancient chants of creation rituals to complex mathematical equations describing primordial energy fluctuations, humankind has always looked for ways of expressing its fascination with the mystery of Creation. In this book we have shared the awe of mythmakers and scientists alike, as they tried to bridge being and becoming with their stories and theories. We have traced the long path leading from myth to science, a path embellished by the many tales of courage and despair, failure and success of those who helped shape our ever-changing view of the Universe. Hopefully, I have given you a glimpse of what is known, and of what is not known or knowable.

As we have seen, cosmology is the only branch of physics that deals with questions asked outside science. That causes cosmology, and cosmologists, to be perceived somewhat differently from other scientists. (Biologists studying the origin of life have similar social roles.) The general public will read far fewer books about the properties of magnetic materials or lasers than they will about the Universe, even though their everyday lives are much more dependent on magnetic materials and lasers than on ideas coming from the big-bang model or black holes.

Certainly, many cosmologists are committed atheists. They don't (and shouldn't!) look for God or any religious connection in their equations or data. Still, they are attracted by the "grand" questions, ranging from the origin of the Universe and the origin of matter to the distribution of galaxies in the Universe at large. It would be very foolish of me to try to understand why certain physicists decide to become cosmologists; their reasons will be as varied and diverse as there are cosmologists across the world. We are the product of our choices, and what we choose to do with our lives is as subjective as it can be. But at least I can speak for myself.

In my case, the decision to become a cosmologist was inspired by the classic popular book by Steven Weinberg, *The First Three Minutes*. I was a junior at the Catholic University of Rio and had to present a seminar on a topic of my choice, when I became completely fascinated by Weinberg's book. I learned that there was a scientific way to approach questions of origin, to the point of having a model, the big-bang model, that actually makes quantitative predictions about the first moments of existence of the Universe. And even more spectacularly, some predictions were actually verified by experiments! Cosmology was no hocus-pocus, it was a quantitative science. I realized with tremendous excitement that it was possible to be a physicist and still work on these "grand" questions. Inspired by Einstein's "cosmic religious feeling," I decided that this was the only path for me. I could actually get paid to think about the origin of the Universe or the origin of matter! (And to teach, of course.)

The more I learned about relativity, quantum mechanics, and how they are applied to the study of cosmology, the more I wanted to learn. And as usual, the more you learn, the more you realize how little you know, how limited we are when facing the infinite creative power of nature. Science is a process, it has often been said. I would add that science is an endless process, that we will never reach an end, simply because there is no end. Whenever I hear pronouncements claiming "the end of science," asserting that all great discoveries that should have been made have already been made, I shudder with disbelief. Can people be so blind to history and to our vast ignorance? Just think of Laplace's "supermind," or the state of confidence of many late-nineteenth-century physicists, and how completely wrong and taken aback they were in their illusions. I wonder how much of this confidence of having reached an end is an expression of unrealized dreams and fantasies.

Nature will never cease to surprise and to amaze us. Our theories of today, of which we are justifiably proud, will be child's play for future generations of scientists. Our models of today will be poor approximations to future models. And yet, the work of future scientists will not be possible without ours, just as ours would not have been possible without Kepler's or Galileo's or Newton's. Science is never completely "right," scientific theories are never the final truth. They evolve and change, get corrected and more efficient, but are never finished. Strange new phenomena will always defy our imagination, those we weren't expecting or couldn't have predicted. We will scramble to understand the new, just as we have ever done. And through this endless pursuit, we will continue to make sense of ourselves and of the world around us, just as we have ever done.

To a smaller or larger extent, we all take part in this adventure; we all share in the rapture of discovery, if not by being directly involved in research, then at least by understanding the ideas of those who expand our human boundaries through their creativity. In this sense, you, me, Heraclitus, Copernicus, and Einstein are all partners in the rhythmic dance of the Universe. It is the persistence of the mysterious that moves us on.

GLOSSARY OF SELECTED TERMS

ABSOLUTE SPACE: According to Newtonian physics, the geometric arena where physical phenomena take place. Its properties are independent of the state of motion of observers.

ABSOLUTE TIME: According to Newtonian physics, absolute time flows at the same precise rate, independent of the state of motion of observers.

ACTION AT A DISTANCE: The assumption, essential in Newtonian physics, that bodies may influence each other without mutual contact, as in the case of the gravitational attraction between the Sun and the planets.

AETHER (Aristotelian): The material substance of heavenly bodies above the lunary sphere.

AETHER (Electromagnetic): The medium believed by nineteenth-century physicists to support the propagation of electromagnetic waves.

ATOMISM: A doctrine originated by pre-Socratic philosophy in ancient Greece where the Universe is composed of small, indivisible components called atoms.

BINDING ENERGY: The energy associated with the binding, by the action of an attractive force, of two or more components in a physical system. It is also the energy released by the system when its bonds are broken by the action of an external agent, or by some intrinsic instability of the system.

BLACKBODY: A perfect absorber of radiation. Kirchhoff showed that the interior of a hollow cavity could mimic a perfect blackbody. In order to study the nature of the radiation produced by heating the walls of the cavity, Kirchhoff made a small hole in one of the walls. The radiation is known as blackbody radiation and is uniquely determined by the temperature of the blackbody.

CALORIC HYPOTHESIS: The hypothesis that heat is an invisible fluid that is transferred between bodies on contact. The caloric hypothesis was abandoned in 1789, after the detailed studies of Benjamin Thompson.

CENTRIPETAL FORCE: A force acting in the direction of the center of motion.

COMPLEMENTARITY PRINCIPLE: Introduced by Bohr, it states that wave and particle are complementary and mutually exclusive ways of representing quantum objects.

COSMIC RAYS: Showers of high-energy particles that pour from outer space into our atmosphere.

COSMOGONY: The study of the origin of the Universe.

COSMOLOGICAL CONSTANT: A parameter, introduced by Einstein in 1917, to guarantee the stability of his finite and static model of the Universe.

COSMOLOGICAL PRINCIPLE: Introduced by Einstein, it states that the Universe is on average the same everywhere and in all directions. Mathematically, it translates into saying that the Universe is homogeneous and isotropic. This principle was extended in 1948 by the proponents of the steady-state model to include also no change in time, the so-called "perfect cosmological principle."

COSMOLOGY: The study of the physical properties and evolution of the Universe.

CREATION MYTH: A myth concerned with the creation of the world.

CRITICAL DENSITY: The energy density that determines if the Universe will expand forever or if it will contract into a big crunch. Its value is roughly 10^{-29} g/cm^3.

CURVATURE RADIUS: The time-dependent parameter that determines the relative distance between two observers in homogeneous and isotropic cosmologies.

DECOUPLING: According to the big-bang model, decoupling is the event that took place when atoms formed and photons broke loose from their interactions with protons and electrons, being thus free to roam the Universe. These photons presently have a black-body spectrum at approximately three absolute degrees, the so-called cosmic background radiation detected by Penzias and Wilson in 1965.

DEISM: The belief in a God who created the world and its natural laws, but who does not interfere with the world after that.

DOPPLER EFFECT: The effect where waves from a source in motion relative to an observer have their wavelengths changed. For a source approaching an observer, the wavelength decreases, while for a source receding from an observer, the wavelength increases.

ELECTROMAGNETIC RADIATION: The radiation emitted by accelerated electric charges.

ELECTRON: The negatively charged elementary particle found in atoms.

ENTROPY: A quantitative measure of the degree of disorder of a physical system. According to the second law of thermodynamics, the entropy of an isolated system cannot decrease.

EPICYCLE: A mathematical device developed in ancient Greece to model celestial orbits. It consists of a circle to which a celestial body is attached, which has a center revolving around another larger circle.

EQUIVALENCE PRINCIPLE: Introduced by Einstein, it states that the effects of gravity can be mocked by an accelerated motion.

FIELD: A physical effect that exists in a region of space. The effect is a manifestation of one or more of the four fundamental forces in nature: gravitation, electromagnetism, and the strong and weak nuclear forces.

FORCE: An action on a body capable of changing its state of motion.

FREQUENCY (of wave): The number of wave crests passing by a fixed point per second.

GRAVITATIONAL REDSHIFT: The increase in the wavelength of radiation emitted in a gravitational field.

GROUND STATE: The lowest energy level of a physical system. For quantum systems, the ground state energy is never zero due to Heisenberg's uncertainty principle.

HUBBLE'S LAW: The relation, obtained by Hubble in 1929, that distant galaxies are receding from us with a velocity proportional to their distance. The relation represents the expansion of the Universe.

INDUCTION: A process by which a moving magnet can generate an electric current in a nearby electric circuit.

INERTIA: The response of a body to any change in its state of motion.

INERTIAL MOTION: Motion with constant velocity with respect to a fixed point or background. It is indistinguishable from being at rest with respect to the same fixed point or background. A body in inertial motion will remain so unless acted upon by an external force (Newton's first law).

INTRINSIC LUMINOSITY: The intrinsic luminosity of an object is the total energy emitted per time. The apparent luminosity of an object falls with distance according to the inverse-square law, that is, with the square of the distance between object and observer.

ISOTOPE: A chemical element is identified by the number of protons in its nucleus; atoms with the same number of protons in their nuclei but different number of neutrons are called isotopes.

KINETIC ENERGY: The energy carried by moving bodies.

KINETIC THEORY: The branch of physics that studies the thermal properties of systems assuming they are composed of small components.

LINES OF FORCE: A visualization technique developed by Faraday to spatially represent the effect of electric and magnetic fields.

MASS: A measure of the bulk quantity of matter of a body.

MOMENTUM: The product of a body's mass with its velocity.

NON-EUCLIDEAN GEOMETRIES: Geometries of curved spaces, as opposed to Euclid's flat space geometry.

NUCLEOSYNTHESIS: The process by which heavier atomic nuclei are synthesized from lighter ones, in a buildup process. Primordial nucleosynthesis refers to the period in the early Universe when the lighter nuclei were synthesized. Stellar nucleosynthesis refers to the synthesis of heavier nuclei during the last stages of a star's life.

NUCLEUS: The positively charged component of atoms. It consists of protons and neutrons bound together by the strong nuclear force.

OLBERS'S PARADOX: The question as to why, in an infinite Universe with an infinite number of stars, the night sky is dark and not infinitely bright.

PHLOGISTON: Hypothetical fluid believed prior (and during) the time of Lavoisier to be released during the burning of substances.

PHOTOELECTRIC EFFECT: The effect where electromagnetic radiation of short wavelength (e.g., violet or ultraviolet) can electrify a piece of metal by kicking electrons away from the surface. Einstein's interpretation of the effect in terms of photons earned him the Nobel prize in 1921.

PHOTON: Light exhibits the so-called wave-particle duality, behaving as a wave or particle depending on the experimental setup. In scattering phenomena, light behaves as particles with energies proportional to their frequencies. These "bundles of light" are called photons.

POTENTIAL ENERGY: Energy stored in a system. For example, a spring can store potential elastic energy, while a body lifted to a certain height stores gravitational potential energy.

PRE-SOCRATIC PHILOSOPHY: Greek philosophy covering the period between Thales of Miletus in the sixth century B.C. to roughly the time of Socrates's birth in 470 B.C.

PRINCIPLE OF RELATIVITY: The laws of physics are identical for all inertial observers. The constraint of inertial motion is removed by the general theory of relativity, which relates the force of gravity to the curvature of spacetime.

QUANTUM VACUUM: The lowest energy state of a quantum system.

QUARKS: The proposed constituents of protons, neutrons, and all other particles that interact via the strong force. There are currently six identified types (or flavors) of quarks—up, down, strange, charm, bottom, and top—all observed in the laboratory.

RATIONAL MYSTICISM: A term I introduced to express the essentially religious inspiration that plays a part in the creative process of many scientists past and present.

REFLECTOR TELESCOPE: A telescope whose principal focusing element is a mirror.

REFRACTION: Deflection from a straight path by a light ray going from one medium (e.g., air) to another (e.g., water).

REFRACTOR TELESCOPE: A telescope whose principal focusing element is a lens.

RETROGRADE MOTION: The apparent backward motion of planets when viewed against the background of stars.

SAVING THE PHENOMENA: As prescribed by Plato, the effort to reduce the complicated motions of the heavenly bodies to simple circular motions.

SINGULARITY: In its technical use, a term expressing the breakdown of mathematical expressions describing the behavior of an otherwise continuous function at a particular point.

SPACETIME: According to the theory of relativity, spacetime is the four-dimensional arena where physical events take place. Distances in spacetime are preserved under relative motion.

SPECTROSCOPE: A device that separates light into its different components.

SPECTRUM: The different components of electromagnetic radiation separated by some device. For example, the spectrum of visible light is composed of the seven colors of the rainbow.

STANDING WAVE: A standing wave is composed of two or more overlapping waves propagating in opposite directions in such a way that the resulting wave does not appear to move.

STEADY-STATE: The condition of no apparent change achieved dynamically through an exact compensation of loss and gain.

STELLAR PARALLAX: The apparent motion of nearby stars against the background of distant stars as viewed by terrestrial observers.

STRONG FORCE: The interaction, active at nuclear distances, responsible for keeping the atomic nucleus together. It is roughly one hundred times more powerful than the electric repulsion of protons in the nucleus.

SUPERNOVA: The explosive event marking the death of a very massive star, during which it may become one billion times brighter than the Sun.

THEISM: The belief in the existence of a God or gods whose presence is immanent in the world.

THERMODYNAMICS: The branch of physics that deals with properties of heat based on macroscopic quantities and measurements.

UNCERTAINTY PRINCIPLE: In its most popular form, Heisenberg's famous principle states that we cannot know with arbitrary precision both the position and momentum of a particle.

WAVELENGTH: The distance between two consecutive crests of a wave.

WAVE-PARTICLE DUALITY: Electromagnetic radiation and matter can exhibit both particlelike and wavelike behavior, depending on the experimental setup. The wave-particle duality is only relevant for small objects whose behavior is dictated by quantum mechanics, such as atoms and subatomic particles.

WEIGHT: The force on a mass due to gravitation.

BIBLIOGRAPHY AND
FURTHER READING

In this Bibliography you will find all works cited in the text as well as additional works of interest. Many of the original works can be found in the collection *Great Books of the Western World*, Mortimer J. Adler, editor in chief. Another important source of original works is the compilation edited by Milton Munitz, *Theories of the Universe*. For myths I was greatly influenced by B. Sproul's *Primal Myths*.

Adler, Mortimer J., ed. *Great Books of the Western World.* Chicago, IL: Encyclopaedia Britannica, 1990.

Barnes, Jonathan. *The Presocratic Philosophers.* London, England: Routledge & Kegan Paul, 1979.

Barrow, John D. *The Origin of the Universe.* New York, NY: Basic Books, 1994.

———and Joseph Silk. *The Left Hand of Creation: The Origin and Evolution of the Expanding Universe.* Oxford, England: Oxford University Press, 1983.

Berger, A., ed. *The Big Bang and Georges Lemaître.* Dordrecht, Holland: D. Reidel, 1984.

Born, Max. *The Born-Einstein Letters.* New York, NY: Walker, 1971.

Clark, Ronald W. *Einstein: The Life and Times.* London, England: Hodder and Stoughton, 1973.

Cohen, I. Bernard. *Revolution in Science.* Cambridge, MA: Harvard University Press, 1985.

——and Richard S. Westfall, eds. *Newton: Texts, Backgrounds, Commentaries.* New York, NY: Norton, 1995.

Cohen, M. R. and I. E. Drabkin. *A Source Book in Greek Science.* Cambridge, MA: Harvard University Press, 1975.

Christianson, Gale E. *Hubble: Mariner of the Nebulae.* New York, NY: Farrar, Straus & Giroux, 1995.

Crowe, Michael, J. *Modern Theories of the Universe.* New York, NY: Dover, 1994.

Davies, Paul C. W. *The Mind of God: The Scientific Basis for a Rational World.* New York, NY: Simon & Schuster, 1992.

De Santillana, Giorgio. *The Crime of Galileo.* Chicago, IL: University of Chicago Press, 1955.

Dictionary of Scientific Biography. New York, NY: Scribner, 1972.

Dobbs, Betty Jo Teeter. *The Janus Face of Genius: The Role of Alchemy in Newton's Thought.* Cambridge, England: Cambridge University Press, 1991.

Dyson, Freeman. *Disturbing the Universe.* New York, NY: Harper & Row, 1979.

——. *Infinite in All Directions.* New York, NY: Harper & Row, 1988.

Einstein, Albert. *The Meaning of Relativity.* Princeton, NJ: Princeton University Press, 1956.

——. *Autobiographical Notes.* Paul Arthur Schilpp, trans. and ed. La Salle, IL: Open Court, 1979.

——. *Ideas and Opinions.* Carl Seelig, ed. New York, NY: Crown, 1982.

Ferris, Timothy. *Coming of Age in the Milky Way.* New York, NY: Morrow, 1988.

Feynman, Richard P., Robert B. Leighton, and Matthew Sands. *The Feynman Lectures on Physics.* Reading, MA: Addison-Wesley, 1963.

Freund, Philip. *Myths of Creation.* New York, NY: Washington Square Press, 1966.

Gamow, George. *The Creation of the Universe.* New York, NY: Viking Press, 1952.

Glashow, Sheldon L. *From Alchemy to Quarks.* Pacific Grove, CA: Brooks/Cole, 1994.

Harrison, Edward. *Masks of the Universe.* New York, NY: Macmillan, 1985.

Hetherington, Norriss S., ed. *Cosmology: Historical, Literary, Philosophical, Religious, and Scientific Perspectives.* New York, NY: Garland, 1993.

Holton, Gerald. *Thematic Origins of Scientific Thought.* Cambridge, MA: Harvard University Press, 1973.

Kirk, G. S., J. E. Raven, and M. Schofield. *The Presocratic Philosophers: A Critical History with a Selection of Texts.* Cambridge, England: Cambridge University Press, 1983.

Koestler, Arthur. *The Sleepwalkers.* Middlesex, England: Penguin, 1959.

Kolb, Rocky. *Blind Watchers of the Sky.* New York, NY: Addison-Wesley, 1996.

Kundera, Milan. *The Unbearable Lightness of Being.* London, England: Faber and Faber, 1984.

Leach, Maria. *The Beginning: Creation Myths Around the World.* New York, NY: Funk & Wagnalls, 1956.

Lederman, Leon M. and David N. Schramm. *From Quarks to the Cosmos: Tools of Discovery.* New York, NY: Freeman, 1989.

Lindley, David. *The End of Physics: The Myth of a Unified Theory.* New York, NY: Basic Books, 1993.

Lloyd, G. E. R. *Early Greek Science: Thales to Aristotle.* London, England: Chatto & Windus, 1970.

Long, Charles H. *Alpha: The Myths of Creation.* New York, NY: Braziller, 1963.

Maclagan, David. *Creation Myths: Man's Introduction to the World.* London, England: Thames and Hudson, 1977.

March, Robert. *Physics for Poets.* New York, NY: McGraw-Hill, 1992.

Misner, C. W., K. S. Thorne, and J. A. Wheeler. *Gravitation.* New York, NY: Freeman, 1973.

Moore, W. *Schrödinger: Life and Thought.* Cambridge, England: Cambridge University Press, 1989.

Motz, Lloyd and J. H. Weaver. *The Story of Physics.* New York, NY: Avon, 1989.

Munitz, Milton K., ed. *Theories of the Universe: From Babylonian Myth to Modern Science.* Glencoe, IL: Free Press, 1957.

North, John. *The Norton History of Astronomy and Cosmology.* New York, NY: Norton, 1995.

Oppenheimer, Julius Robert. *Science and Common Understanding.* New York, NY: Simon & Schuster, 1953.

Pagels, Heinz R. *The Cosmic Code.* New York, NY: Bantam Books, 1982.

——. *Perfect Symmetry.* New York, NY: Bantam Books, 1986.

Pais, Abraham. *Subtle Is the Lord: The Science and the Life of Albert Einstein.* Oxford, England: Oxford University Press, 1982.

Park, David. *The How and the Why: An Essay on the Origins and Development of Physical Theory.* Princeton, NJ: Princeton University Press, 1988.

Rhodes, Richard. *The Making of the Atomic Bomb.* New York, NY: Simon & Schuster, 1988.

Silk, Joseph. *The Big Bang: The Creation and Evolution of the Universe.* San Francisco, CA: Freeman, 1980.

Smith, Robert W. *The Expanding Universe: Astronomy's "Great Debate" 1900–1931.* Cambridge, England: Cambridge University Press, 1982.

Smoot, George and Keay Davidson. *Wrinkles in Time.* New York, NY: Morrow, 1993.

Sproul, Barbara C. *Primal Myths.* San Francisco, CA: Harper San Francisco, 1979.

Stoppard, Tom. *Arcadia.* London, England: Faber and Faber, 1993.

Taub, Liba Chaia. *Ptolemy's Universe.* Chicago, IL: Open Court, 1993.

Thorne, Kip S. *Black Holes and Time Warps: Einstein's Outrageous Legacy.* New York, NY: Norton, 1994.

Weinberg, Steven. *The First Three Minutes.* New York, NY: Basic Books, 1977.

Wells, H. G. *A Short History of the World.* Middlesex, England: Penguin, 1965.

Westfall, Richard S. *Never at Rest: A Biography of Isaac Newton.* Cambridge, England: Cambridge University Press, 1980.

——. *Essays on the Trial of Galileo.* Notre Dame, IN: Notre Dame University Press, 1989.

Whitehead, Alfred North. *Science and the Modern World.* New York, NY: Macmillan. 1925.

Zee, A. *Fearful Symmetry: The Search for Beauty in Modern Physics.* New York, NY: Collier, 1986.

——. *An Old Man's Toy: Gravity at Work and Play in Einstein's Universe.* New York, NY: Macmillan, 1989.

INDEX